Environmental Earth Sciences

Series editor

James W. LaMoreaux, Tuscaloosa, USA

For further volumes:
http://www.springer.com/series/8394

Werner Balderer · Adam Porowski
Hussein Idris · James W. LaMoreaux
Editors

Thermal and Mineral Waters

Origin, Properties and Applications

Springer

Editors
Werner Balderer
Department of Earth Sciences
Schweizerische Geotechnische Kommission
Zürich
Switzerland

Adam Porowski
Warszawa
Poland

Hussein Idris
Cairo
Egypt

James W. LaMoreaux
PELA GeoEnvironmental, Inc.
Tuscaloosa, AL
USA

H. Idris is deceased

ISBN 978-3-642-28823-4 ISBN 978-3-642-28824-1 (eBook)
DOI 10.1007/978-3-642-28824-1
Springer Heidelberg New York Dordrecht London

Library of Congress Control Number: 2014934304

Printed on acid-free paper

Springer is part of Springer Science+Business Media (www.springer.com)

Dedication of this volume of contributions from the CMTW 2009 Egyptian Meeting (October 10–18, 2009)

To the memory of Eng. Hussein Idris, Vice Chairman of CMTW

Almost from the beginning of the International Association of Hydrogeologists (IAH) Commission on Mineral and Thermal Water (CMTW) in 1968 Eng. Hussein Idris was a very active member of the Commission

Under his leadership, two great meetings were held in Egypt in 2004 and 2009. The 2009 meeting was organized as a Special Session to review 60 years of Eng. Idris' professional activity in groundwater exploration, engineering, and management

The theme of the 2009 CMTW meeting was: Groundwater, Thermal and Mineral Water in Areas of Arid conditions: Consequences for the Current Situation of Climate Change and the Increasing Population of Egypt. The following topics were examined:

Upconing of saline waters in urban areas

Hydrogeology of the Sinai Desert

Possibilities for the use of hot waters in desert areas as sources of energy

Future development of oil fields: use of oilfield waters for balneological purposes and as a source of raw minerals (as trace elements) and organic materials

There were also contributions by members of the CMTW and of Egyptian scientists related to other topics. During the meeting, tours focused on the following aspects of Egypt's geology:

Tour 1: Trip to historical sites of Cairo: Coptic Monastery, First Coptic churches in Cairo, first Mosque and first Synagogue, Sadat Memorial and outlook to Citadelle

Tour 2: Trip from Cairo to the Sinai desert related to the hydrogeology of arid areas and the utilization of hot waters in the desert; a visit to hot water well Nr. 20; the thermal spring of Bath of Pharaoh (Hammam Farun); and the Monastery of St. Catherine

Tour 3: Groundwater problems around Cairo: The upconing of saline groundwater emerging in surficial ponds as environmental threats to the fast growing city

Tour 4: Groundwater and petroleum production: visits to Qarun Oilfield and the Baharia Oasis, with its deep groundwater wells in the Nubian Sandstone

The two meetings of 2004 and 2009 were unforgettable as is the memory of Eng. Idris, a wise man, engineer, and practitioner in hydrogeology. He oversaw the installation of more than 100 deep groundwater wells in Egypt and Libya. He was not only a great geologist, but he was also a great personality, who served Egypt as a Colonel during the Sinai wars and as a diplomatic arbiter in an international conflict between Sudan and Yugoslavia in the 1970s

Eng. Idris served as Deputy Chairman and Joint Managing Director of Algulf, Ltd. of

Cairo, Egypt, and London, England. He was a consultant on short-term assignments in the Sudan for the Food and Agriculture Organization (FAO) of the United Nations and for the Arab Agricultural Development Organization of the Arab League. Previously he served as Chairman of the Land Reclamation and Agricultural Development Authority, was President of The General Company for Research and Ground Water (REGWA, the only Egyptian company specializing in research and groundwater). Throughout his career, he worked for the Egyptian government in various capacities. As Resident Irrigation Engineer of the Ministry of Public Works, he was responsible for the groundwater development in the Sinai, and Kharga and Dakhla Oases where he authorized well drilling and supervised a detailed hydrogeologic study

This unforgettable man died on January 2, 2013, but his memory will live on in the history of the CMTW, as demonstrated by the following publication: "Western Desert of Egypt: Genie of the Desert." Genie of the Desert was Eng. Idris' nickname and it was appropriate because he knew the desert like the back of his hand. Eng. Idris also prepared the "Annotated Bibliography on Mineral and Thermal Waters." The bibliography, with about 1,500 references, stands as a major achievement of the CMTW

Eng. Idris was well respected by his peers as shown in a few of their remarks below:

"He was dear, concerned, caring, and had an active, modern mind and thoughts"— Magda Idris;

"One of the most respected members, pillar of annual meetings, wise voice"—Adam Porowski;
"Open dialogue and common benefits; contributed to success and viability of the Commission"—Peter Kralj;
"Through these unforgettable excursions Idris helped us discover and fall in love with his wonderful country"—Natalia Vinograd;
"Often his remarks were combined with a wink of fine humor; and even though his body was weak in later years, there was a young mind open to everyone with a lot of life's experiences grown on the soil of esprit"—Kurt Von Storch;
"He and my father (Dr. Philip LaMoreaux) began working together in the Western Desert of Egypt in the 1950s. He was like a second father to me over the years and he certainly served a similar role for all of us and the CMTW. He will be sorely missed"—Jim LaMoreaux

Organization Committee

Organizers

- The Arab Mining and Petroleum Association (AMPA)
- Egyptian National Committee of the International Association of Hydrogeologists (IAH)
- Egyptian Society of Engineers
- General Company for Research and Groundwater "REGWA"

Organizing and Scientific Committee

Shawki Abdin	Petroleum Geologist, member of the IAH Commission on Mineral and Thermal Waters (CMTW), member of the American Association of Petroleum Geologists (AAPG), board member of the Geological Society of Egypt and Arab Mining and Petroleum Association (AMPA)
Ismail Moftah	Engineer, member of the IAH, member of the AMPA, member of Society of Petroleum (SP), member of the American Chamber of Commerce, and Chairman of Engineering Consultancy Office for Oil and Gas in Egypt
Magda Idris	Hydrogeologist, member of the IAH, member of the AMPA
Ahmed M. R. Khater	Professor, member of the IAH, Director of Research Institute for Groundwater (RIGW)
Abd El Meguid Amer	Professor, Faculty of Engineering Azhar University, member of the AMPA, member of the Management Engineering Society, member of the Egyptian Engineering Society, member of the Association of Egyptians Graduated from German Speaking Institutions
Mohamed El Wagih	Professor, Head of The Mining and Petroleum Engineering Department Cairo University, member of the AMPA, member of the Egyptian Engineering Society

Ibrahim Zahran	Petroleum geologist
Hussein Idris	(Deceased) Hydrogeologist, Deputy Chairman of the IAH CMTW, member of the AMPA
Dr. Jim LaMoreaux	Chairman of the IAH CMTW, Chairman of PELA GeoEnvironment, Inc.
Dr. Adam Porowski	Hydrogeologist, Secretary of the IAH CMTW, Hydrogeologist at the Institute of Geological Sciences Polish Academy of Sciences
Dr. Werner Balderer	Hydrogeologist, Deputy Chairman of the IAH CMTW, Department of Earth Sciences of ETH Zurich, member of the Swiss Geotechnical Commission (SGTK)

Contents

Introduction

This special volume contains selected papers from the 38th Meeting of the Commission on Mineral and Thermal Waters of the International Association of Hydrogeologists (IAH-CMTW) held in Cairo, Egypt, on 10–18 October 2009. The meeting was organized by Engineer Dr. Hussein Idris, Deputy Chairman of the Commission, with the support of the Egyptian Chamber of Engineers and the Petroleum Society of Egypt. During scientific sessions at the meeting, several lectures were presented by members of the Commission as well as invited lecturers, local authorities, academic staff, and management staff from the petroleum and mining industry.

Some of the most interesting lectures are presented herein to demonstrate the variety of mineral and thermal waters occurrence and utilization in different countries of the world. Four papers concern mineral and thermal waters of Egypt. A detailed overview of the geology and thermal water potential of the Sinai Peninsula is given by Mohamed Ragaie El Tahlawi, Professor from the Mining and Metallurgy Department, Assiut University. His paper focuses on manifestations of thermal waters of outflow temperature between 35 and 72 °C at famous areas such as Oyun Mousa, Ras Sidr, Hammam Pharaun, and Hammam Mousa. Ibrahim El Shamy, Professor from the Hydrogeology Department, Helwan University, presents an interesting compilation concerning the mineral and thermal water springs in Helwan district. The important and urgent problems of pollution and mineralized groundwater ascension in the karstic area of southeastern Cairo of El Emmam El Shaffei district is presented by engineers of the REGWA Co., namely Aly Werwer, Hamed Shaker, and Magda Idris. Their presentation focuses on three old springs: Ain Al Sira, Khayalat Al Shorta, and Abo Al Saud, which are used to form big ponds in the city. Since the pond of Abo Al Saud spring was filled up with rocks and wastes and replaced finally by a garden, the water level started to rise in the remaining ponds. Nowadays the water level is dramatically high and frequently causes flooding in the adjacent district of Cairo. The team of Werner Balderer, in collaboration with Hussein Idris, made additional research concerning the origin of waters of these ponds. It appears that the waters are outflows of the outcropping Tertiary rocks in close vicinity, as well as outflows of outcropping Eocene formations of the East banks of the river Nile. Dr. Balderer and colleagues from Switzerland present another significant problem of high fluoride concentrations in drinking groundwater in several villages of Dogubeyazit area, Eastern

Fig. 1 Main point of interest of the field trip during CMTW IAH meeting in Cairo, 2009. *1*, Ain Al Sira, Cairo; *2*, Khayalat Al Shorta, Cairo; *3*, Helwan district; *4*, Baharia Oasis; *5*, Ayun Mousa; *6*, Hammam Faraun; *7*, Mount Sinai. Map prepared based on Google Earth data

Anatolia, Turkey. Hydrogeological and hydrochemical investigations performed in the spring of 2002 revealed fluoride concentration in drinking water well above the WHO recommended limit of 1 ppm. Moreover, in three villages the fluoride concentration in drinking water was found in the range of 6–8 ppm, causing severe dental and health problems for the local population. This study discusses alleviating the problem by the following: (i) survey of the existing situation with regard to the current use of groundwater provided by the existing springs and boreholes, (ii) proposal of a new distribution network of groundwater resources with low fluoride concentration to improve the health and quality of life of the affected population.

Another interesting and cutting-edge study is presented by a team of CMTW members, i.e., József Deák and István Fórizs of the Institute for Geological and Geochemical Research of the Hungarian Academy of Sciences, Hungary, Werner Balderer, and coworker Fanny Leuenberger from Eidgenössische Technische Hochschule (ETH), Zurich. This study concerns the application of ^{36}Cl chlorine isotopes to investigate the origin of thermal waters in the area of Budapest, Hungary. Based on the ^{36}Cl isotopes the thermal waters extracted from springs and wells in Budapest are shown to have two components of meteoric origin. One component is connected with the modern hydrological cycle and contains tritium and ^{36}Cl typical for anthropogenic origin. The second component, the high temperature one, is the deep circulating ancient meteoric water with no tritium and $^{36}Cl/Cl_{total}$ ratio, indicating a very long transit time of groundwater or mixing of such an old component along the flow regime.

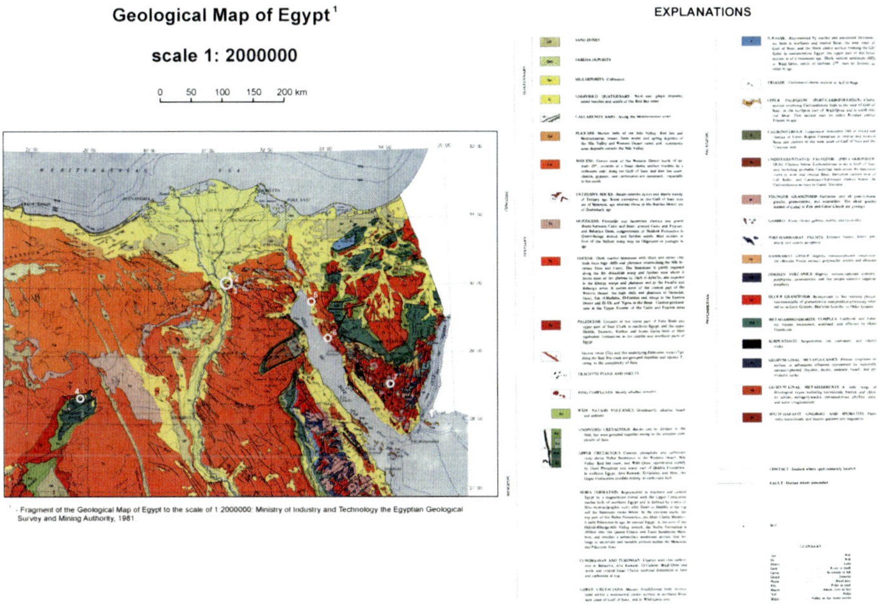

[1] - Fragment of the Geological Map of Egypt to the scale of 1:2000000: Ministry of Industry and Technology the Egyptian Geological Survey and Mining Authority, 1981

For the first time during the CMTW forum a contribution concerning the Crimea Peninsula, Ukraine, was presented. Elena Kayukova, Professor from the Faculty of Hydrogeology, University of St. Petersburg, Russia, provides a comprehensive description of the salt lakes in the Crimea Peninsula, their origin, chemical composition of waters, as well as chemical composition and curative features of the muds from these lakes. The muds formed in some lakes are treated as curative deposits and the resources are geologically recognized and documented.

Contributions from members of the Commission from Poland, Russia, and Australia are included. Adam Porowski, Professor from the Institute of Geological Sciences of Polish Academy of Sciences, Poland, presents a general overview of the chemical and isotopic characteristics of thermal waters in Carpathian Geothermal Province, Southern Poland. One of the phenomena in this region is the occurrence of the so-called dehydration waters, which are usually associated with oil and gas deposits. Natalia Vinograd, from the Faculty of Hydrogeology University of St. Petersburg, Russia, shows interesting and valuable research concerning the industrially valuable chemical components that can be recovered from oil field waters in Russia. Andrew Shugg from Sinclair Knight Merz, Melbourne, Australia, presents a very interesting comprehensive overview of high bicarbonate groundwater occurrence in the state of Victoria.

Hussein Idris organized the meeting and led an excellent field trip of the Western Desert. He passed away before this book was published. In his memory, it is dedicated to him and to all the lives he touched, particularly those of the CMTW members.

Saline Groundwater Ascension Problems in East Cairo: Environmental Threats of Fast Growing Cities

Aly Werwer, Hamed Shaker and Magda Idris

Abstract The spread of urbanization increases water demand to satisfy both domestic and industrial aspects. In return, large volumes of wastewater will accumulate. Careful environmental studies before major urbanization projects are very important to minimize hazards due to environmental threats. The southeastern part of Cairo is one of these examples, as large amounts of wastewater were accumulated and leaked down to the ponds of Ain Al Sira, causing degradation to one of the oldest therapeutic and skin curing locations.

Keywords Urbanization · Environmental threats · Water demand · Egypt

1 Introduction

Ain Al Sira, Khayalat Al Shorta and Abo Al Saud springs in the southeastern part of Cairo, Egypt's famous district of El Emmam El Shaffei used to form three main ponds. Three decades ago, one of them (Abo Al Saud) was filled with whatever dump was available. A housing project was constructed around the filled pond, while a public garden replaced the filled pond. Nowadays, the area occupied by the two main ponds (Ain Al Sira and Khayalat Al Shorta) is more than 100,000 m^2 (Fig. 1). The water surface of the two ponds is increasing tremendously and flooding the nearby area roads and cemeteries.

Ain al Sira pond which was well known among Egyptians since a long time ago for its therapeutic and curing effects, as both its water and clay were used for many skin disorders and diseases, is now polluted with sewage water to the limit that

---

A. Werwer · H. Shaker · M. Idris (✉)
REGWA Company, Cairo, Egypt
e-mail: magda2731@gmail.com

W. Balderer et al. (eds.), *Thermal and Mineral Waters*, Environmental Earth Sciences,
DOI: 10.1007/978-3-642-28824-1_1,

Fig. 1 General location of the two ponds from Google Earth

made the governmental medical authority prohibit its therapeutic use. Another unrealized hazard is the damage that can happen in case the karst limestone develops sinkholes.

2 Position of the Study Area

The study area is bounded on its western side by the River Nile and by the escarpment of El Mokattem Hills on the eastern side. It is characterized by low relief with general slope from south to north and from east to west. It is essentially occupied by sedimentary rocks ranging in age from Quaternary to Tertiary. The study area occupies a part of the regional homocline structure developed on the foreland of the great Arabo-Nubian Massive, leading northward into the Mediterranean geosynclinal basin. This region is bounded on the eastern and western sides, respectively, by two taphro-geosynclinal basins of the Suez Gulf and of the Nile Delta.

3 Hydrogeological Settings

In the vicinity of Cairo, groundwater forms two main aquifer systems, namely: Quaternary and Tertiary. Both aquifer systems remain in hydraulic contact.

3.1 Quaternary Aquifers

Holocene Aquifer is composed of Nile silt, clay and sand and ranges in thickness between 10 and 14 m, where the main recharge is the direct downward infiltration from irrigation channels.

Pleistocene Aquifer is composed of silt, graded sand and gravels with total thickness ranges between 50 and 80 m, where the main recharge is the vertical water seepage from irrigation channels and the river Nile (Attia 1954).

3.2 Tertiary Aquifers

Pliocene aquifers are mainly composed of fine to medium sands, clay and calcareous and argillaceous sediments. The main source of recharge within the study area is upward leakage from Eocene fractured limestone and the downward seepage is from irrigation channels and drains.

Eocene aquifer or fractured limestone has a wide distribution on the surface and in the subsurface of the Helwan district. Abdel Daiem (1971) expected that the ideal Eocene sections within the area range between 400 and 800 m and overlay the unconformable Cretaceous Nubian Sandstone aquifer. Secondary porosity is more common than primary porosity. Eocene fractured limestone aquifer displays karst features, which lead to the formation of several springs within the study area. These springs are recharged by upward water leakage from the underlying major Nubian Sandstone aquifer through fissures, crakes and fault planes.

Springs feeding Ain Al Sira pond are emerging from fissures belonging to Eocene limestone (El Ramly 1969). However, there were different opinions regarding the genesis of groundwater emerging through the fissures of Eocene limestone in the area east of Cairo. Gastinel (1862) and Azadian (1930) reported groundwater in Ain Al Sira is due to direct infiltration from the River Nile. Ramadan et al. (1960) considered that it is recharged from groundwater invading basement and volcanic rocks. Hazzaa et al. (1966) regards that it is due to seepage at the foothills. Hemida and Abd El-Daem (1976) were not in favor of direct seepage of Nile water or even Pleistocene water, as they consider that the source can be the deep groundwater leaking from Lower Cretaceous Nubian Sandstone.

The increase of water consumption in the study area due to the urbanization expansion of Cairo that took place in the area as well as in Mokattam and neighboring housing developments, plus the consecutive sewage systems caused considerable leakage, adding a new recharge source for Ain Al Sira and Khayalat Al Shorta springs. In other words, the housing development that mushroomed lately in the study area without taking into consideration the limited capacity of the available infrastructure caused serious damage to the water supply and sewage pipes, causing considerable leakage that added to the sources of recharge.

4 Chemical Characteristics of Groundwater

4.1 Ain Al Sira Pond

It forms a large pond occupying a depression within Eocene limestone (Figs. 2, 3). The analysis of water samples collected during 2009 proved that it is highly saline water with total dissolved solids of about 51,440 ppm, which can be represented by the following hydro chemical formula (Eq. 1):

$$M^{51.4}\frac{Cl^{95}SO_4^4HCO_3^1}{(Na+K)^{82}Mg^{11}Ca^2} \qquad (1)$$

where, M—total dissolved solids; main ions in % meqv, pH of the water measured in the field is 7.9.

Hence, it is a sodium chloride water type, and the water temperature in the pond ranges from 21 °C in January up to 25 °C according to seasonal variation. However, Gastinel (1862) reported a temperature of 40 °C at the bottom of the pond when the air temperature was about 26 °C, which indicates that groundwater is feeding the pool at a temperature higher than what is actually measured.

4.1.1 Recent Conditions of Ain Al Sira Pond

Figures 2 and 3 indicate the increase of the surface of the pond. Hi increase can be due to the large volumes of wastewater. Also, the dilution in the total dissolved salts compared to old records of 114,500–187,000 ppm confirms leakage of large quantities of water from the housing developments (Hemida and Abd El-Daem 1976).

4.2 Khayalat Al Shorta Pond

It is located at about 600 m to the south of Ain al Sira, with a water level of 20 m above sea level as measured in 2009 (Figs. 4, 5). Depth of water in the pond varies from one point to another as water salinity during the study was 8,240 ppm. Water temperature ranges between 21 °C in January and 26 °C in July matching seasonal variations of air temperature with a chemical composition represented by the following formula (Eq. 2)

$$M^{8.2}\frac{Cl^{76}SO_4^{20}HCO_3^3}{(Na+K)^{60}Mg^{17}Ca^{23}} \qquad (2)$$

where, M—total dissolved solids; main ions in % meqv, pH of the water measured in the field is 7.9.

Fig. 2 Ain Al Sira Pond on 9 July 2000

Fig. 3 Ain Al Sira Pond on 20 February 2008

4.2.1 Recent Conditions of Khayalat Al Shorta Pond

Figures 4 and 5 indicate that the increase in the surface area of the Khayalat Al Shorta Pond is increasing. This increase can be due to the increasing volumes of wastewater in the nearby urban district, especially Mokattam City, Basateen and Ezbet Shenod. The salinity of the pond is decreasing compared to the old records of 7,960–180,000 ppm due to water seepage (Hemida and Abd El-Daem 1976).

Fig. 4 Khayalat Al Shorta Pond on 9 July 2000

Fig. 5 Khayalat Al Shorta Pond on 20 February 2008

5 Conclusions

As the population is growing rapidly, urbanization spreads vigorously causing an increase in the water demand to satisfy different needs for both domestic and industrial aspects. In return, tremendous volumes of wastewater are seeping from the old sewer systems, which are running full and under pressure, causing considerable leakages that ultimately add to the water in the ponds. Hundreds of new buildings were constructed around the banks of the two ponds as well without actual developing or adaptation of the sewage infrastructure casing gradation for the whole surroundings. Meanwhile the dumping of one of the three ponds (Abo Al Saud) decreased the evaporation surface. Hence, the discharge was decreased. At the same time the rapid growth of Cairo increased the recharge of the limestone. Obviously, those are good reasons for raising the level of the water in the ponds and hence flooding the surrounding area.

Such a situation may explain the increasing surface area for both lakes Ain Al Sira and Khayalat Al Shorta Pond, and the dilution noticed on their chemical composition.

Much attention should be considered for any future urbanization expansion to prevent severe water quality issues that would emerge from inadequate treatment of wastewater, and special attention must be given to increase public awareness to change water consumption patterns.

Authorities responsible for developing housing schemes should take into consideration that filling these ponds or part of it with dump to create new areas for the housing schemes will definitely raise the level of the groundwater and flood the vicinity. It is a must to maintain sufficient surface area for the ponds to evaporate a quantity of water equal to the recharge. This can be assured by considering the geological and hydrogeological conditions of a project location within the visibility study of new housing projects.

References

Abdel Daiem AA (1971) Hydrogeological studies of springs in the area to the East of Cairo. MSc thesis, Faculty of Science, Ain Shams University, p 120

Attia MI (1954) Deposits in the Nile Valley and the Delta Mines and quarries. Dep Geol Surv Egypt, p 356

Azadian A (1930) Les Eaux d'Egypte, Egypte, Min Int., Le Caire (Notes at Rapport des Laboratories de l'Hygiene Publique)

El Ramly IM (1969) Recent review of investigation on the thermal and mineral springs. In: XXIII International Geological Congress, vol 19, pp 201–213

Gastinel M (1862) Me'moire sur les eaux salines froides de Ain-Syra, pres Le Caire: Mem. Inst. Eg., le Caire, I

Hazzaa IB, Raef M, El Sabban MZ, Wahbi F (1966) Determination of radioactivity and chemical constituents in water springs in U.A.R. J Egypt Med Assoc 47, paper no. 2/3

Hemida IH, Abd El-Daem AA (1976) Hydrogeochemical characteristics of some thermal mineral springs in the area to the East of Cairo Egypt-A.R.E. In: International congress on thermal waters, geothermal energy and volcanism of the mediterranean area, Athens, Oct 1976

Ramadan FM, El-Hinnawi EE, Swelim AA, Moussa RS (1960) Evaluation of Ain El Sira ("Ein Elsera") Spring waters. J Egypt Med Assoc 43:950–958

Origin, Anthropogenic and Climate Influences on the Occurrences of Saline Groundwater at the City of Cairo, Egypt Deduced by Chemical Parameters of the Water Composition

Werner Balderer, Fanny Leuenberger and Hussein Idris

Abstract Three old springs in the southeastern part at the border of Cairo City, named Ain Al Sira, Khayalat Al Shorta and Abo El Soud, used to be the origin of three natural ponds. Since the removal of one of these natural ponds (Abo El Soud) about three decades ago, by filling it with stones and recycled building material, an ongoing rise of the water level of the two remaining lakes or ponds has been observed. The water surface rise of the two ponds is still continuing and is actually even flooding nearby roads and cemeteries. The origin of the groundwater is from outflows of the outcropping Tertiary rocks according to the main chemical composition of Pond Nr. 1 Khayalat Al Shorta, as also outflows of the water of the Eocene formation outcrop along the Eastern bank of the Nile valley according to the main chemical composition of Pond Nr. 2 Ain Al Sira. A remediation of the present situation would most probably only be possible by restoring the pools to their original extent and maintaining the water level by controlled drainage of the groundwater overflow, by installing a drainage system for lowering the groundwater level with small pumping stations scattered throughout the area of high groundwater levels along the main streets in the direction of the Nile (which would possibly replace the disappeared historical channel system).

Keywords Water logging · Cairo ponds · Remediation of groundwater level · Origin and chemical composition of waters of Cairo ponds

H. Idris is deceased

W. Balderer (✉) · F. Leuenberger
ETH Zurich, Zurich, Switzerland
e-mail: werner.balderer@erdw.ethz.ch

W. Balderer et al. (eds.), *Thermal and Mineral Waters*, Environmental Earth Sciences,
DOI: 10.1007/978-3-642-28824-1_2,

1 Introduction

Three old springs in the southeastern part at the border of Cairo City, named Ain Al Sira, Khayalat Al Shorta and Abo El Soud, used to be the origin of three lakes (see: Werwer et al. this volume).

Since the removal of one spring (Abo El Soud) about three decades ago by filling it with stones and recycled building material, an ongoing rise of the water level of the two remaining lakes or ponds has been observed. As indicated in Werwer et al. (this volume), the area of the two main ponds (Ain Al Sira and Khayalat Al Shorta) today occupies more than 100,000 m^2. The water surface rise of the two ponds is still continuing and actually even flooding nearby roads and cemeteries.

Before a remediation of this very ugly situation of overflowing lakes can be planned and carried out, the origin of this water inflow to these lakes by the springs involved (fed by groundwater) has to be identified. The aim of this study was to elucidate the origin of the water of these two remaining lakes/ponds, named Khayalat Al Shorta (Pond Nr. 1) and Ain Al Sira (Pond Nr. 2). Along with the observed increase in water level, the marked increase in salinity of these highly saline ponds must still be monitored.

Within this context the following possibilities of origin of the spring/lake water (according to the hydrogeological situation as described in Attia (1999) and Tahani (1979) will be further discussed based on the chemical composition of these waters.

By seepage from irrigation channels (water origin from Nile).

From the karstified carbonate aquifer system, assigned to the Eocene and to the Upper Cretaceous, which outcrops along the Nile system:

1. By upcoming of saline deep water of Nubian Sandstone, outcropping in the Western Desert;
2. By outflow of water from the outcropping fissured and weathered hard rock aquifer system, assigned to the Pre-Cambrian, outcrops in the Eastern Desert and the Sinai Peninsula;
3. By local groundwater aquifer in gravel to sandy sediments with a thickness of up to 60 m confined or semi-confined at the top by a thin layer of clay;
4. By uncontrolled sewage release from growing new settlements within this area of Cairo, which are infiltrating into the local groundwater;
5. A mixing of these possibilities of water origin.

2 Hydrogeological Situation

According to the geological map of Cairo by Tahani (1979) (Fig. 1), a connection of this local aquifer to the limestones, which are outcropping in the eastern part of Cairo, seems to be most probable. It may even be that local outflows to these two existing ponds occur in connection with the observed faults of East–West direction

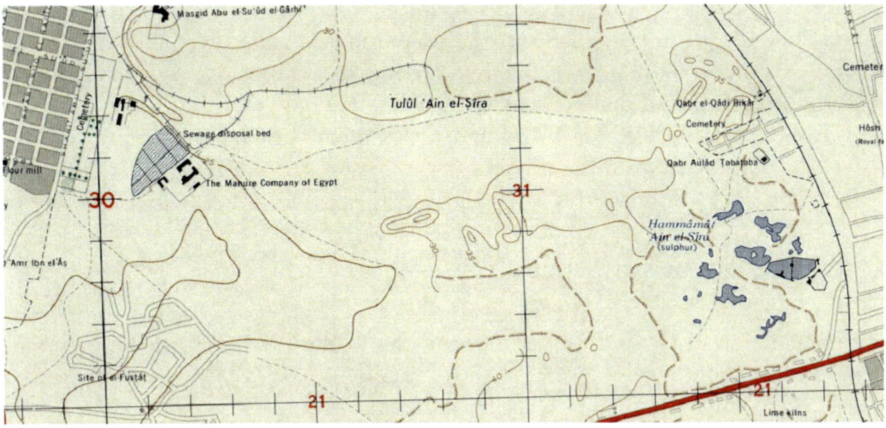

Fig. 1 Pond area in Cairo according to the Map 1:10,000 of the U.S. Army Map Service (1958): Cairo—South Cairo No. 1, Edition 4-AMS, Series P971

(Tahani 1979). However, this hydrogeological hypothesis must also be in agreement with the result of the chemical investigations to be accepted as a valid explanation.

3 History of the Ponds as Part of the Evolution of Cairo City

In the 16th to 18th centuries, canals and ponds outgoing from the Nile were systematically built, for irrigation purposes and as transport for ships. However, these networks of canals and related ponds fulfilled also twofold hydrological, mainly unknown, tasks: (i) first, as a tampering system in the Nile flood events; the network of canals and ponds were able to store temporarily the incoming flood water and in this way prevent the general flooding of Cairo City; (ii) secondly, during lower Nile water levels, it functioned as a drainage system by uptake of the outflow of groundwater and thus maintaining a sustainable groundwater level below the City of Cairo. After the construction of the Aswan High Dam in the 20th century (1970), these canals were no longer needed for irrigation and as storage capacity of the flood water. As a consequence of increased population growth, traffic development and intensified settlements, most of the canals and connected ponds were successively filled and used for road network extensions along former canals and for construction of new buildings above the filled ponds. If these canals, as man-made constructions, were situated above the groundwater level, their filling had no further consequences; the remaining water dried up by evaporation and by downward seepage to deeper soil horizons, as it was no longer fed by flooding Nile water.

However, some parts of the canals, and the ponds of Ain Al Sira, Khayalat Al Shorta and Abo El Soud which were now no longer connected to a canal system showed another contradictory behavior: As these ponds were filled up with dumping materials and one of them removed even completely from the landscape (Abo El Soud), the water was invading the basements of the constructed buildings as well as outflowing in other places.

4 Hydrogeological Explanation

The explanation of these observed phenomena is that these two remaining ponds are related to natural groundwater outflow, regulating its level like an overpressure valve.

4.1 The Hydrogeological Situation can be Traced Back in the Shorter History as Follows

- A view of the lakes in 1958, before the construction of the Aswan High Dam is shown on the topographical map 1:10,000 of the U.S. Army Map Service, 1958: Cairo—South Cairo No. 1, Edition 4-AMS, Series P971 (Fig. 1) where several open water bodies are documented in the area of the remaining ponds of Ain Al Sira and Khayalat Al Shorta.

  This map, also documents a sewage disposal bed (of the Manure Company of Egypt) situated only about 1 km west of the Ain el Sira pond area.

- In an initial study after the construction of the Aswan High Dam (finished in 1970) the effect of the rise of the groundwater level in the area of the city of Cairo was already observed and described by Youssef Tahani in 1979.
- The geological setting of the area of Cairo is presented in detail in Werwer et al. (this volume) as also in Tahani (1979) and Attia (1999).
- The main rock units, which according to Tahani (1979) (Fig. 2) outcrop in the closer area of Cairo, are: (i) the limestone unit assigned to Eocene and Upper Cretaceous (containing a karstified carbonate aquifer); (ii) covered by old sand dunes; these deposits are in their turn overlain near the eastern shore of the Nile river by; (iii) old gravel terraces; the central and lower areas of Cairo and the whole area towards the north are completely covered by; (iv) alluvium deposits, which are partly covered by; (v) the most recent flood sediments consisting of sand and gravel.

  These alluvium deposits underlying the whole area of Cairo City contain an aquifer in gravel to sandy sediments with a thickness of up to 60 m. As this aquifer is covered by a thin layer of clay it is characterized as a confined to semi-confined aquifer.

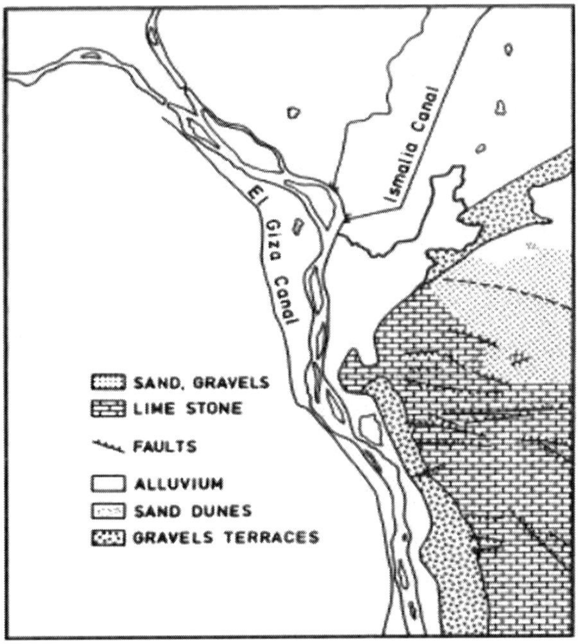

Fig. 2 Geological map of greater Cairo Region (Tahani 1979)

- According to the map of Tahani for the year 1979 (Fig. 3), the groundwater levels within the City of Cairo (situated on the Eastern bank of the river Nile) are in large parts (especially more distant from the River Nile) higher than the regulated Nile level of about 16.5–17 m a.s.l. (Fig. 4), if not situated in the areas of artificially lowered levels and influenced by nearby groundwater pumping stations.
- Therefore, the Nile can be considered as acting as a drainage level for the groundwater below Cairo City according to the map of groundwater levels in 1979 presented in Fig. 3.

5 Results and Interpretation of the Chemical Investigation of the Involved Water Sources

To test all the hypotheses of the origin of the water in the pond Nr. 1 and Nr. 2 in Cairo and to test possibilities of mixing to explain the present-day composition of the lake water the following water samples were analyzed:

- Low mineralized Nile water (origin also of water supply of Cairo municipality)
- Water of spring 1 and 2 in Helwan
- Original high saline groundwater (Sudr well and Hamam El Farun spring in Sinai), as potential end-member of the inferred mixing process
- Water of Alexander well in Baharia (from Nubian Sandstone, Western Dessert)

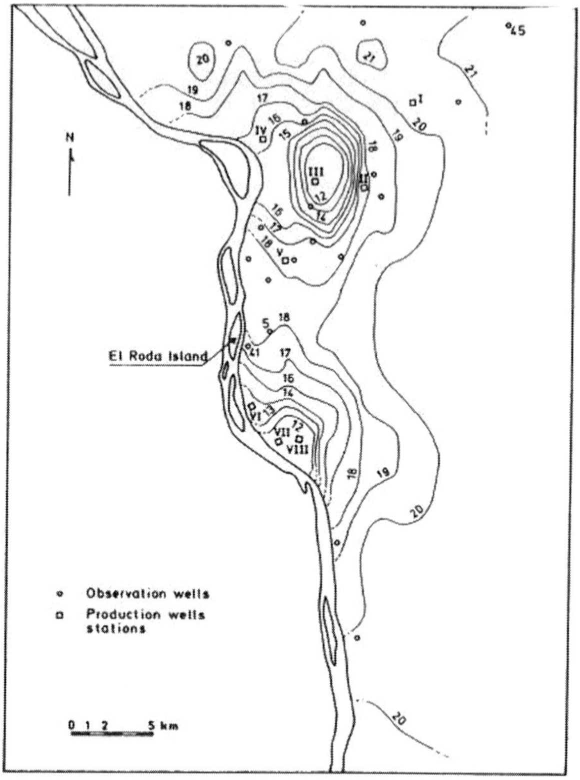

Fig. 3 Map of groundwater levels (water table) for the year 1979 according to Tahani (1979)

The results of anion and cation composition of the analyzed water samples (by Ion chromatography) are presented in Table 1.

6 Discussion of Origin of the Pond Waters 1 and 2 Based on the Chemical Composition of the Analyzed Water Samples by Means of Specific Diagrams

The resulting diagrams of main ionic composition are presented in Figs. 5 and 6 as Schoeller Diagrams (concentrations in meq/l). In addition to the analyzed water samples, the single available analyses of sewage water composition originating from Giza and Alexandria, as reported in Abdel-Ghaffar (1988) as well as a one water analyses from a well named M Nebwi NE 3, which is situated between Sudr and Hamam el Farun, as reported in Issar et al. (1971), were also introduced in the diagrams of the chemical composition for comparison.

Considering the individual groundwater composition curves in the Schoeller type diagram of Fig. 5, the following systematics can be derived:

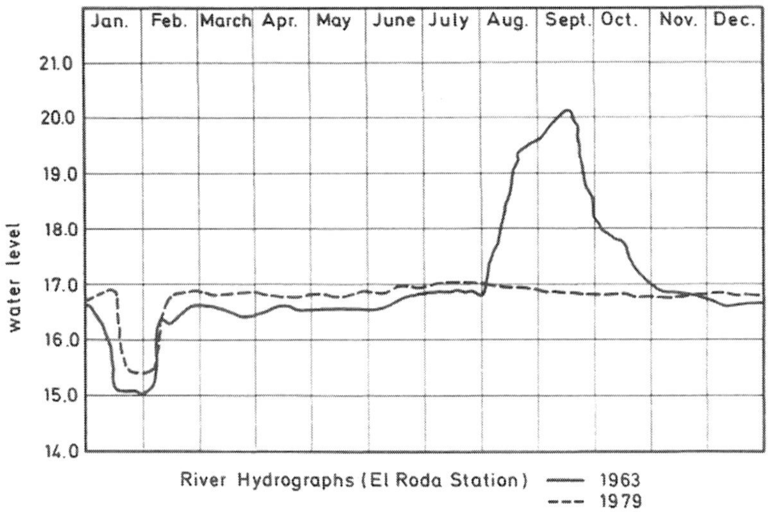

Fig. 4 Water level of Nile River (hydrograph) before 1963 and after operation of Aswan High Dam measured at El Roda Nile Island in 1979

- The waters of the Nile and the groundwater of the Alexander well of Baharia Oasis are much less mineralized than all other investigated waters.
- Considering individual ion concentrations and shape of the curves in Fig. 5, the water of Pond Nr. 2 matches quite closely the curves of Sudr well Nr. 20 of samples taken 28.5.2009 as well as 13.10.2009. A bit less concentrated, but nearly the same shape show the curves of the springs Helwan 1 and Helwan 2.
- Considering the two curves of Pond Nr. 1, which are almost similar, except some difference in the bicarbonate (alkalinity) concentration, they almost coincide with the curve of the historical water analyses of one sample of the M Nebwi NE 3 mentioned in Issar et al. (1971).
- The curves for the two samples of Hamam el Farun are situated in the Schoeller diagram in Fig. 5 between the curves of Pond Nr. 1 and Pond Nr. 2 and may therefore be considered as potential mixing components.

In the Schoeller type diagram of Fig. 6, a selection of analyzed water samples, as well as the uniquely available analytical results of sewage water composition originating from Giza and Alexandria, as reported in Abdel-Ghaffar (1988) are represented. According to this diagram the following relations are now appearing:

- The curve of the water of Pond Nr. 2 quite closely matches the curves of Sudr well Nr. 20 of samples of 13.10.2009 as already mentioned, and those of the sewage type water of Alexandria, as reported in Abdel-Ghaffar (1988).
- The two nearly identical curves of the two samples of Pond Nr. 1 pretty nearly coincide with the historical sample of the M Nebwi NE 3 well of 1971, which is characterized in Isaar et al. (1971) as Miocene formation water.

Table 1 Results of measurements and analyses of the investigated groundwater samples

Sampling date	Location	W. temp.	El. cond. (µS/cm)	pH	Cl (ppm)	NO_2 (ppm)	Br (ppm)	NO_3 (ppm)	SO_4 (ppm)	Na (ppm)	NH_4 (ppm)	K (ppm)	Mg (ppm)	Ca (ppm)	Sr (ppm)
03.06.2009	P1, 1		74700	6.55	26956.00	37.00	33.00		5757.00	15893.89		635.00	1297.00	1425.00	53.03
15.1 0.2009	P1, 2	36	82700	6.36	31177.98		42.02		6920.42	19277.57		754.84	1560.62	1678.90	57.67
03.06.2009	P 2		11190	7.51	2446.00		0.00		3162.00	1570.00		487.00	240.00	790.00	16.12
03.06.2009	Nile		896	8.10	162.00		0.00		53.00	110.00		9.00	17.00	38.00	0.00
15.1 0.2009	Hel1	37	6260	7.29	1599.74		0.00	6.15	725.31	890.97		15.52	144.55	258.84	17.74
15.1 0.2009	Hel2	38	8270	7.06	1950.61		4.99		1441.95	1225.08	3.32	23.01	168.00	405.50	19.30
16.10.2009	Alex Baha	38	454	7.66	53.82		0.00		8.89	35.75	0.36	27.72	16.97	13.30	0.00
28.05.2009	Sudrl		11340	6.55	3404.00		27.00		624.00	1438.00		56.00	136.00	1009.00	13.82
28.05.2009	Far 1		24100	6.44	7676.00	19.00	50.00		714.00	3672.00		113.00	282.00	1127.00	18.79
13.10.2009	Sudr 2	87	11350	6.35	3526.72		26.92		658.67	1510.76		60.76	143.00	715.20	12.94
13.10.2009	Far 2	76	24900	6.20	8674.60		52.63		767.06	3990.63		120.68	304.39	1206.08	19.45

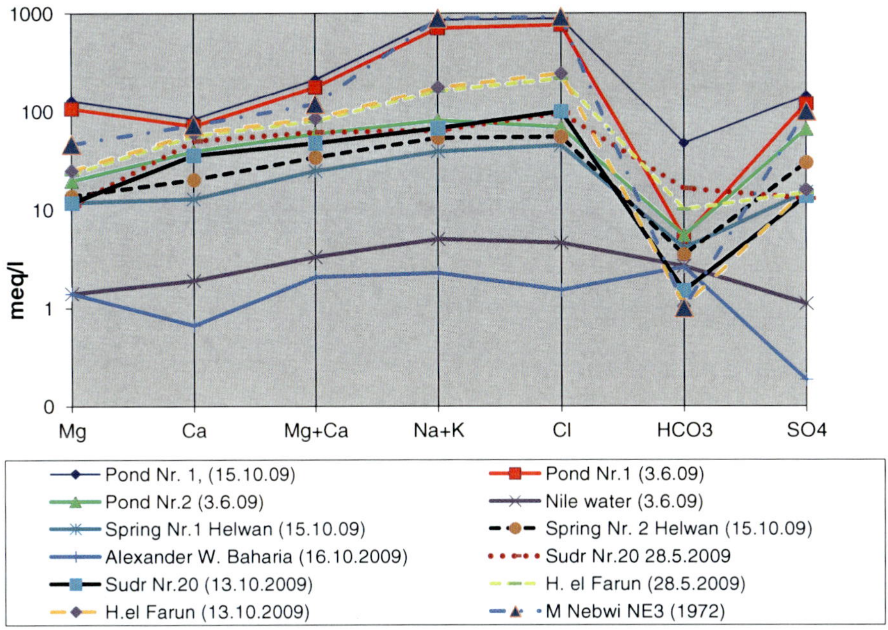

Fig. 5 Schoeller diagram of the investigated waters of Cairo and of the enlarged area for interpretation of origin of Cairo Ponds water

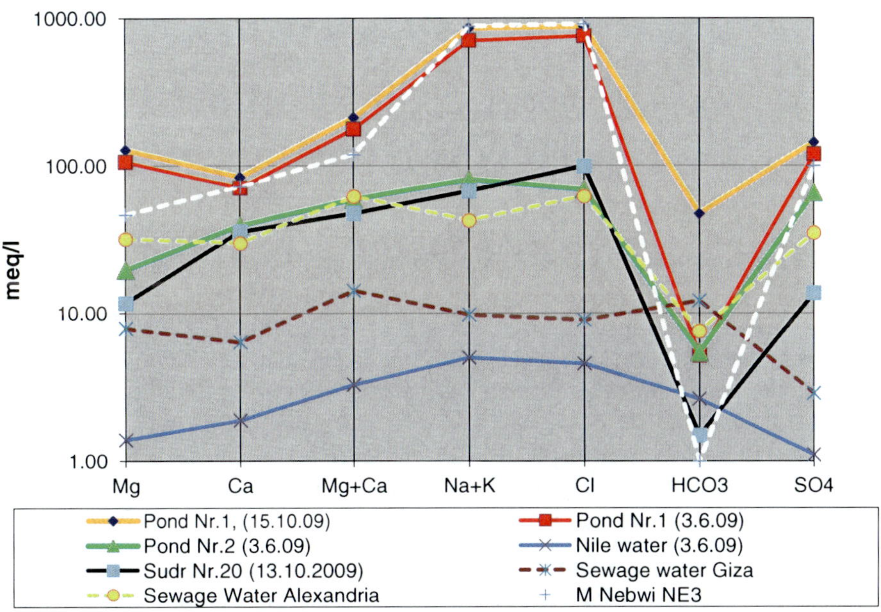

Fig. 6 Schoeller diagram of the investigated waters of Cairo and of enlarged area for interpretation of origin of Cairo Ponds water including analyses of sewage water from Giza and Alexandria (Abdel-Ghaffar 1988) and of the M Nebwi NE 3 well (Issar et al. 1971)

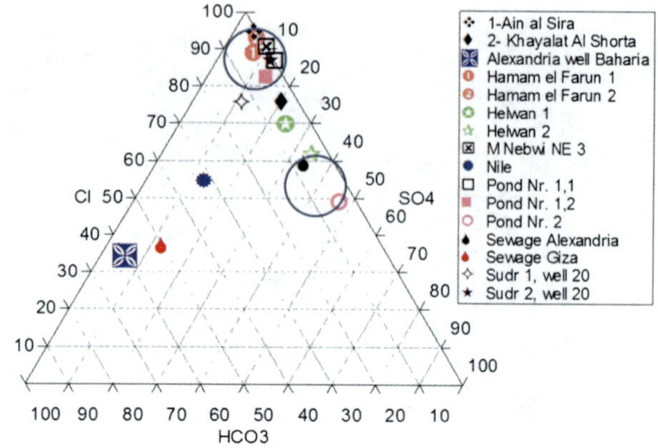

Fig. 7 Main anion composition of the investigated waters represented in a modified Piper diagram

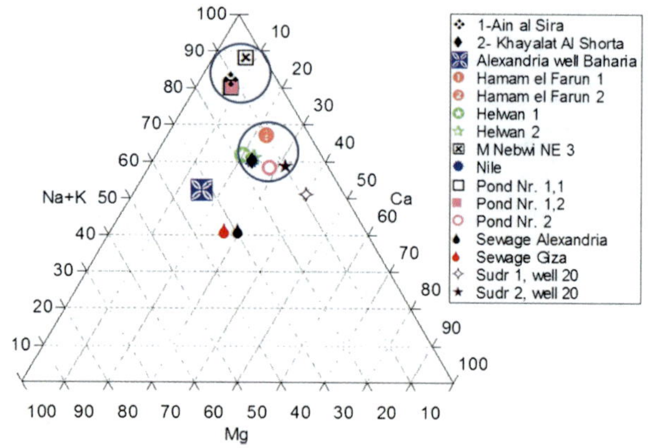

Fig. 8 Main cation composition of the investigated waters represented in a modified Piper diagram

In Figs. 7 and 8, the main ionic composition is represented in piper diagrams (with concentrations in meq % (of total sums of anions and cations). These diagrams of anions and cations include also the ionic compositions of the pond waters of Ain Al Sira (Pond Nr. 1) and Khayalat Al Shorta (Pond Nr. 2) as reported in Werwer et al. (this volume).

In these Piper diagrams of Figs. 7 and 8, the main anion and cation composition is reflected by their ion-equivalent rations (as meq %), where the sum of the main cations (Mg, Ca and Na + K), or anions are each equal to 100 %.

Based on these diagrams the following relationships with respect of the possible origin of the waters of the pond waters of Khayalat Al Shorta (Pond Nr. 1) and Ain Al Sira (Pond Nr. 2) can be stated:

- As in the process of evaporation, no change of the ion ratios occur (as long as no precipitation of dissolved species occurs); the respective position in the Piper diagrams of cations and anions will not change. If this aspect is considered, only points of waters situated at the same place (or very close to each other in the piper diagrams may be considered as potential starting waters for an evolution by evaporation only as would be the case for the higher concentrated waters of pond Nr. 1 and Pond Nr. 2.

- If also the concentrations of the main ions of water samples are very similar, which means that their respective lines nearly coincide in the Schoeller diagram, an almost identical evolution or even the same origin (within flow path) may be considered.
- Therefore, if the data of the investigated waters in the piper diagrams are considered, it results that none of the analyzed waters represent the original starting water for an evolution by only invoking a process of evaporation from the pond itself.

According to the piper diagram of the anions (Fig. 7) the following two clusters (in the circles) can be distinguished:

- One cluster contains the water samples of pond Nr. 1 (Khayalat Al Shorta) as also the water of Hamam El Farun (both samples) and water sample of October 2009 of Sudr well Nr. 20. Additionally, a water analysis from a well named M Nebwi NE 3, which is situated between Sudr and Hamam el Farun as reported in Issar et al. (1971), also fits well within this first cluster.
- The second cluster contains the water sample of October 2009 of Ain Al Sira (Pond Nr. 2), as also the water of Helwan spring 2 (near University) and sewage water of Alexandria.
- Between these two groups on nearly straight line the waters of Sudr well Nr. 20 (sample of May 2009), Helwan spring 1 (in Park Garden) as well as the anion ratios of Ain Al Sira (Pond Nr. 2, as reported in Werwer et al. this volume) are situated. Especially for the Pond Nr. 2 water composition, a mixing process between waters of cluster 1 and 2 may be considered.
- Outside of both clusters, and not grouped are situated the sewage water of Giza, the Nile water and the fresh groundwater of Nubian sandstone of Alexander spring at Baharia Oasis.

According to the piper diagram of cation concentrations (Fig. 8) also two clusters (in the circles) in the domain of sodium-rich waters can be distinguished:

- The main cluster contains the waters of Ain Al Sira (Pond Nr. 2, all analyses), Nile Water, Helwan spring 1 and 2, as well as Sudr well Nr. 20, sample of October 2009, as well as both samples of Haman El Farun.
- The sample of Sudr well Nr. 20 of May 2009 has the same ratio for magnesium as the abovementioned waters, but lower ratios of alkaline and calcium concentrations. It may be considered as diluted by nearby shallow groundwater.
- The groundwater of the Alexander well of Baharia oasis yields higher concentration ratios of magnesium but lower concentrations of calcium and magnesium.
- The two sewage waters yield nearly the same composition of cation concentration ratio of magnesium as the Baharia Oasis water, but lower concentrations

in alkaline and higher calcium concentrations. These sewage waters may therefore be considered as originating from local drinking water supply (of fresh water composition like the Alexander well water) with a contribution of calcium by washing agents.

- A second cluster contains the waters of pond Nr. 1 (Khayalat Al Shorta, all analyses) and one water analyses from a well named M Nebwi NE 3, which is situated between Sudr and Hamam el Farun as reported in Issar et al. (1971).
- This last observation is now very important, as for all three analyses of the water of pond Nr. 1 (Khayalat Al Shorta) and one water from a well named M Nebwi NE 3, a nearly perfect fit of their position in both Piper diagrams of Figs. 7 and 8, for anions and cations results. As the absolute concentrations are quite similar, the water of Pond Nr. 1 may therefore be considered as direct outflow (and originating) of the water of the type of M Nebwi NE 3, which was described by Issar et al. (1971) as Miocene formation waters.

7 Discussion of Origin of the Pond Waters 1 and 2 Based on the Chemical Composition of the Analyzed Water Samples Especially Based on the Resulting Chloride and Bromide Concentrations

According to Table 1 and Fig. 9 the Bromide concentration shows a dependency with the chloride concentration (logarithmic regression coefficient $R^2 = 0.5637$).

The resulting line of logarithmic regression shows a trend from the Nile water as lowest mineralized water on one end and the highest mineralized water of Pond Nr. 1 (sample of October 2009) on the other end. It may, therefore, be interpreted as the dilution linc between two end-members. This means that the outflowing mineralized water of Pond Nr. 1 represents an outflow of deep-seated brine water along existing faults which is mixed with the water of the local gravel aquifer, and which, in turn, is influenced by Nile water and dispersed water of seepage losses of the local water supply.

Similar dilution relations are also revealed by the strontium versus chloride and strontium versus sulphate diagrams in Figs. 10 and 11. The corresponding linear regressions are even more confirmed by coefficients of $R^2 = 0.8697$ and of $R^2 = 0.8697$. In its relationship with respect to Bromide as well as to Strontium, the water of Pond Nr. 1 is confirmed as being the concentrated end-member, which is diluted, compared to all other considered waters.

That may invoke now also the hydraulic process of equilibrium between the water bodies/aquifers of different densities: if the level of the Nile River as the regional discharge system is lowered (decreased) but the hydraulic potential of the dense formation water is constant, the outflow of this deep water in the area of Pond Nr. 1 will increase. The loss of sewage water of the new unregulated settlements will further increase the groundwater level in the local gravel aquifer in

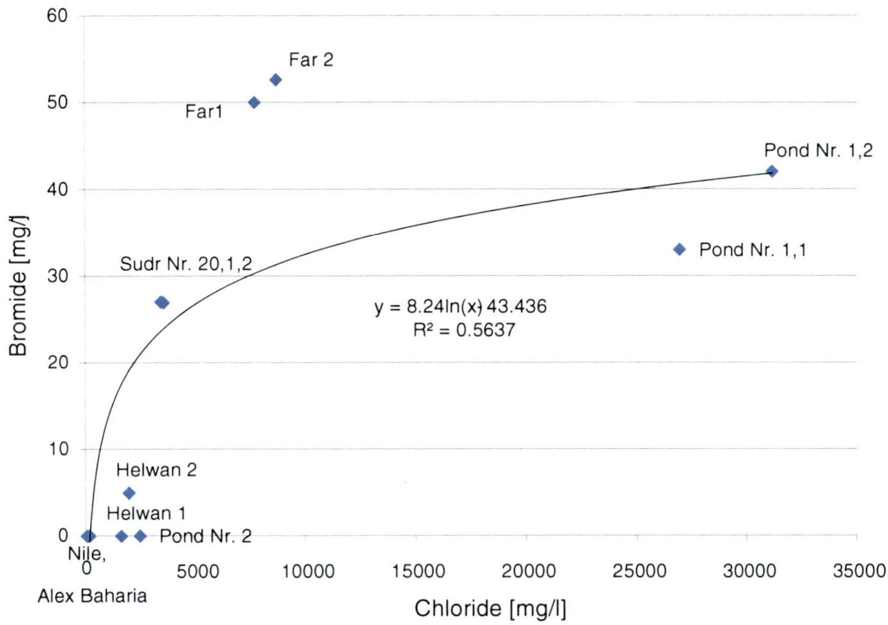

Fig. 9 Bromide versus chloride concentration (mg/l) of the investigated waters

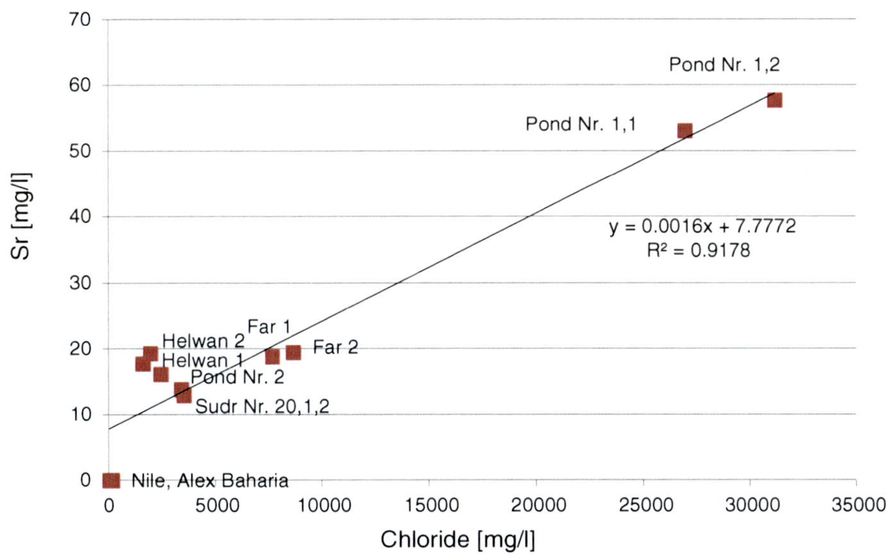

Fig. 10 Strontium versus chloride concentration (mg/l) of the investigated waters

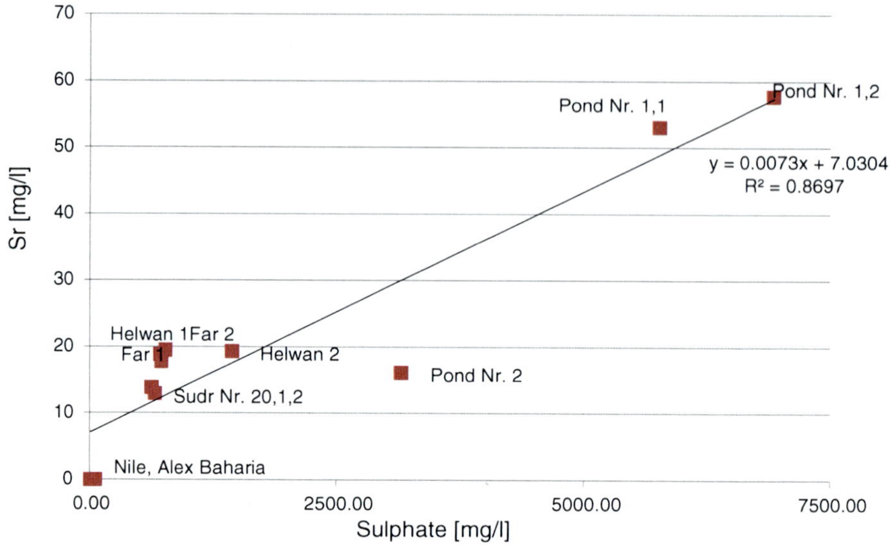

Fig. 11 Strontium versus sulphate concentration (mg/l) of the investigated waters

areas that are not drained anymore by canals connected to the Nile River as they are presently nonexistent.

8 Conclusions with Respect of the Hydrogeological Situation of the Pond Waters of Khayalat Al Shorta (Pond Nr. 1) and Ain Al Sira (Pond Nr. 2)

According to its chemical composition, the highly mineralized water of Pond Nr. 1 Khayalat Al Shorta can reasonably only be derived from outflowing water of the outcropping Tertiary rocks which probably would have a similar chemical composition as the water documented by the water well named M Nebwi NE 3, originating from Miocene formations.

- Concerning the water composition of Pond Nr. 2 Ain Al Sira, an anthropogenic influence of sewage water inflow cannot be ruled out; but the main contribution seems to be natural outflowing groundwater with a composition close to that of the Sudr well Nr. 20 water. This would also imply an outflow from the outcropping Eocene formation near the Nile valley. For both Ponds a flow direction from east to west seems to be most probable, which is also in agreement with the water level map of 1979 as presented by Youssef Tahani.
- The increase of the water level in and around Cairo city, especially on the eastern bank of the Nile already observed in 1979, is most probably due to the infilling of some channels (formerly connected to the Nile River) after the

established High Dam regime. The main effect was that the drainage by the open channels was diminished by their filling up with soils and other materials. This argument is in agreement with the modelled situation by Yossef Tahani. He expressed his findings by the following conclusions:

- To lower the groundwater table, four pumping stations were suggested, each having a steady pumping rate of 40,000 m^3/day.
- This suggested solution for lowering the groundwater table by adding some pumping stations to those already existing was proved by the simulated model to be unsuccessful in lowering the water table.
- The other suggestion of scattering small pumping stations in order to get a uniform sinking of the water table is only a theoretical idea of the author.

If the main problem is the general rise of the water table (groundwater level) in large parts of the city of Cairo, a remediation would only be possible by:

1. Restoring the pools in their original extent, maintaining the water level by controlled drainage of the groundwater overflow,
2. Installing a drainage system for lowering the groundwater table along the main streets in the direction to the Nile (which would possibly replace the disappeared channel system).

References

Abdel-Ghaffar AS (1988) Wastewater treatment and use in agriculture. In: FAO irrigation and drainage paper 47. FAO Natural Resources Management and Environment Department, Food and Agriculture Organization of the United Nations, Rome, 1992, M-56 ISBN 92-5-103135-5

Attia FAR (1999) Water and development in Greater Cairo (Egypt). Afers Internacionals, núm. 45–46:81–102

Issar A, Rosenthal E, Eckstein Y, Bogoch R (1971) Formation waters, hot springs and mineralization phenomena along the eastern shore of the Gulf of Suez. Bulletin of the International Association of Scientific Hydrology, XVI, 3 9/1971

Tahani Y (1979) Control of ground water levels in Cairo, Egypt, after the High Dam. Department of Civil Engineering, Mansoura University, Mansoura, Egypt

Werwer A, Shaker H, Idris M (this volume) Saline groundwater ascension problems in East Cairo: environmental threats of fast growing cities. Springer, Environmental Earth Sciences Book Series

Sinai Peninsula: An Overview of Geology and Thermal Groundwater Potentialities

Mohamed Ragaie El Tahlawi

Abstract The Sinai Peninsula is the Asiatic part of Egypt, covering an area of some 61,000 km^2. It has a triangular shape and is separated geographically from the Eastern Desert by the Gulf of Suez. Unconformable overlying the Precambrian basement is a section of Carboniferous through Eocene sedimentary rocks (shales, sandstones and limestones). Porous sandstone of Upper Jurassic to Lower Cretaceous age, ranging in thickness from 90 to 150 m, is widespread in the northern Gulf area. This sandstone, commonly referred to as the Nubian sandstone, has hydrologic aquifer properties and is overlain by a sequence of Upper Cretaceous through Upper Eocene limestones, dolostones, chalks and marls, which act as an aquiclude to the underlying Nubian sandstone. Sinai area is characterized by superficial thermal manifestations represented by a cluster of hot springs with varying temperatures of 35–72 °C. These springs mostly issue out of fractures and are distributed along the eastern shore of the Gulf of Suez. Four important thermal water therapeutic sites in Sinai, namely Oyun Mousa, Ras Sidr, Hammam Pharaoun and Hammam Mousa, which are located on the eastern shore of the Gulf of Suez, have been described.

Keywords Sinai Peninsula · Groundwaters · Thermal waters · Springs · Geothermal potential

1 Introduction

The Sinai Peninsula is the Asiatic part of Egypt, covering an area of some 61,000 km^2. It has a triangular shape and is separated geographically from the Eastern Desert by the Gulf of Suez (Fig. 1).

M. R. El Tahlawi (✉)
President of the Arab Mining and Petroleum Association, Mining and Metallurgy Department, Assiut University, Assiut, Egypt
e-mail: eltahlawi@yahoo.com

W. Balderer et al. (eds.), *Thermal and Mineral Waters*, Environmental Earth Sciences,
DOI: 10.1007/978-3-642-28824-1_3, © Springer-Verlag Berlin Heidelberg 2014

Fig. 1 Roads map of Sinai Peninsula showing main famous places

Big seismic activity and the tremendous eruptive phenomena have given Sinai its characteristic looks. These activities are controlled by the tectonic activity of the Red Sea, Gulf of Suez and Gulf of Aqaba.

The highest mountains are the Gebel Musa (Moses' mountain) with 2,285 m, and the Sinai's highest Gebel Kathrina (Mount St. Catherine mountain) with 2,642 m. Many of the Pharaohs obtained their precious stones from the southern Sinai.

Sinai area is characterized by superficial thermal manifestations represented by a cluster of hot springs with varying temperatures 35–72 °C. These springs mostly issue out of fractures and distributed along the eastern shore of the Gulf of Suez.

The geology of Sinai has been comprehensively studied, especially after the discovery of oil and many other mineral deposits.

Special attention has been given to the study of the groundwater geothermometry, which is restricted to the surface hot springs. Recent studies on the geothermal activity characterized the geothermal setting of Sinai Peninsula comprising the estimation of the depths to the Curie-point isotherm based on spectral analysis of the aeromagnetic data. Moreover, DC resistivity surveys were conducted at three localities near thermal hot springs to elucidate the possible aquifer (El-Qady et al. 2005).

2 Geological Settings

Geomorphologically, Sinai Peninsula comprises the following seven geomorphic districts:

1. Southern elevated mountains district
2. Central Plateau district comprising
   (a) El Egma questa to the southwest
   (b) El Tieh questa to the north.
3. Hilly district, northeast of the Peninsula, gently sloping to the northeast
4. Coastal district of gently undulated surface marked by sandstone ridges
5. Muddy and marshy land district occupying the shorelines and some lakes.

2.1 Stratigraphy of the Gulf of Suez

The stratigraphy of the Gulf of Suez has been reviewed by many authors (Reynolds 1979; Said 1990, etc.). Unconformable overlying the Precambrian basement is a section of Carboniferous through Eocene sedimentary rocks (shales, sandstones and limestones). Porous sandstone of Upper Jurassic to Lower Cretaceous age, ranging in thickness from 90 to 150 m, is widespread in the northern Gulf area. This sandstone, commonly referred to as the Nubian sandstone, has hydrologic aquifer properties and is overlain by a sequence of Upper Cretaceous through Upper Eocene limestones, dolostones, chalks and marls, which act as an aquiclude to the underlying Nubian sandstone. A major unconformity separates the post-Eocene sediments (syn-rift succession) from the underlying pre-Eocene (pre-rift succession). Miocene sedimentation in the Gulf of Suez was triggered by the Mediterranean transgression over a graben complex. Two main facies were deposited, namely: (1) a clastic phase representing the initial filling up of the basin, and (2) a late evaporitic phase that marked the restriction of the basin.

Evaporites in the northern part of the Gulf of Suez area are almost exclusively made of anhydrite and gypsum and lack halite. Evaporite thickens to the south and towards the center of the trough, reaching up to 610 m in thickness in Ras Lagia (in the vicinity of Hammam Faraoun). These stratigraphic relationships are shown in Fig. 2, a generalized stratigraphic column for the Gulf of Suez that was modified from Schütz (1994).

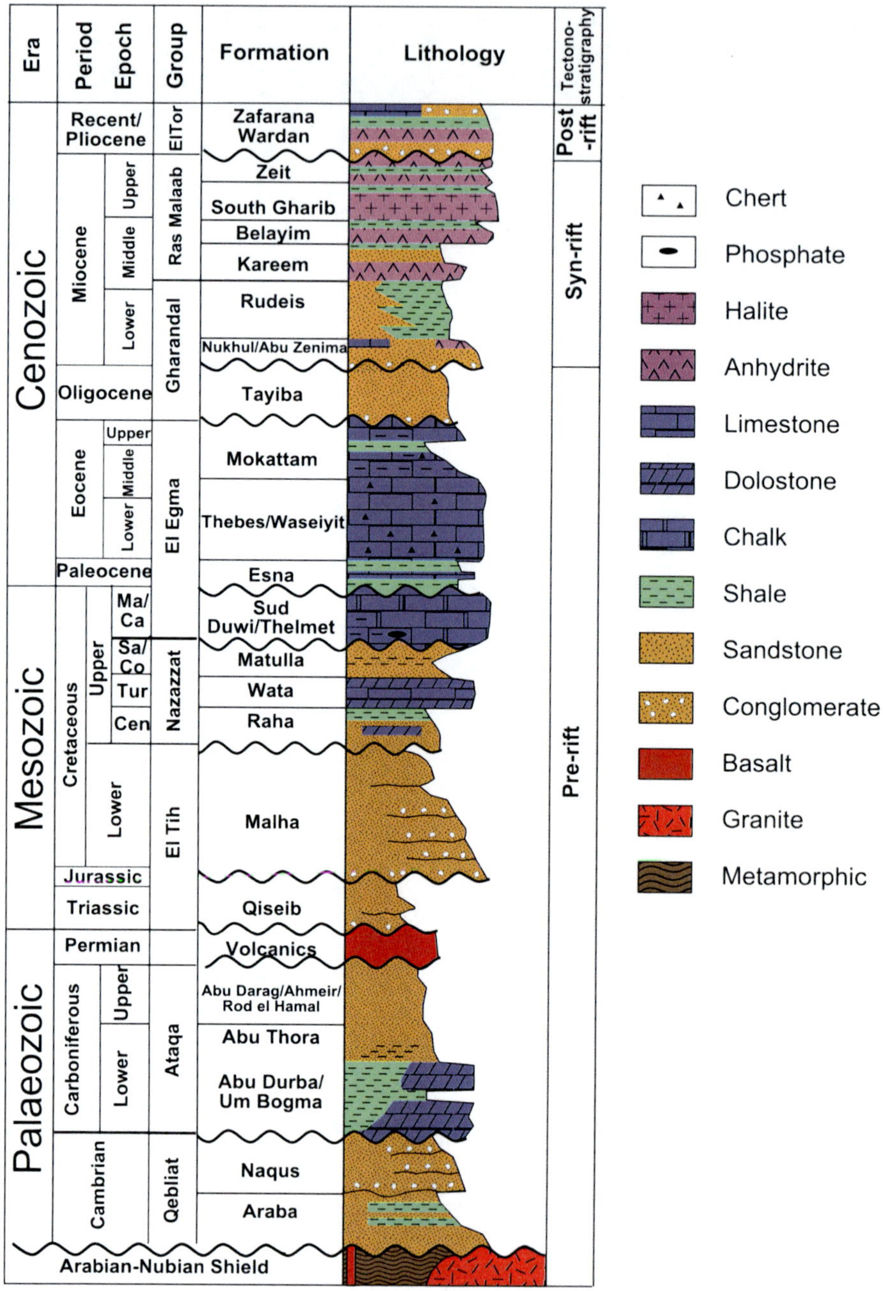

Fig. 2 Generalized stratigraphic column for the Gulf of Suez (modified after Pivnik et al. 2003)

The Peninsula of Sinai may be divided into three geological districts, namely, the granitic and metamorphic, limestone, and sandstone rocks of which they are composed. Generally, Sinai reflects all geologic columns of Egypt. The northern part of Sinai mainly consists of sandstone plains and hills. The Tih Plateau forms the boundary between the northern area and the southern mountainous with towering peaks.

The mass of the rocks in the Sinai Peninsula is cut by amazing valleys, the so-called wadis. They are mostly parallel and can be dozens of kilometers long while hundreds of meters wide. Wadis basically are fossil beds of rivers that run through the region in the Quaternary Era (some 1.8 millon years ago). At that time there was much more rainfall which deeply eroded the rock beds.

The surface geologic units at Sinai are presented in Fig. 3. The Paleozoic rocks are overlying the Precambrian basement in the southwestern Sinai. The Mesozoic strata crops out in the north where an almost complete sequence from Triassic to Cretaceous is known, while it composes a subsurface section attaining a large thickness (955 m for Jurassic rocks only at Oyun Musa).

The stratigraphic column described in detail by Said (1962) is divided into four major units:

1. Upper Calcareous Division, composed of sandstone, shales, conglomerates, evaporites, and foraminiferal marl. They range in age from Oligocene to Pleistocene.
2. Middle Calcareous Division, mainly marls, chalks, limestones and shales of Cenomanian to Eocene age.
3. Lower Clastic Division, in the south; these are Carboniferous to Lower Cretaceous sandstones. Toward the north, the rocks are more calcareous. This complex, as especially its upper part, is also known as the Nubian Sandstone.
4. Basement Complex, Precambrian igneous and metamorphic rocks; many dykes occur crossing the basement rocks and are considered of Tertiary age. Oil is produced along the shore of the Gulf of Suez. Production is mainly from Miocene sandstones, and in certain fields, from Carboniferous, Cretaceous and Eocene formations.

The Nubian Sandstone of Lower Cretaceous represent the main water bearing unit in the region attaining a maximum thickness of about 500 m; at central Sinai it is made up of 70–130 m (Kora 1995).

Among the Cenozoic rocks, the Oligocene is the most important from the geothermal point of view. The end of the Oligocene Period witnessed the rifting movements that brought the Gulf of Suez to its modern shape. Along the numerous faults that crossed the region, simatic molten material climbed up and appeared in the form of sporadic, but frequent, lava flows, sills, sheets and dyke feeders chiefly of olivine basaltic and olivine diabase.

Fig. 3 Simplified geological map of Sinai. Explanations: *1* Pre-Cambrian, *2* Carboniferous, *3* Lower Cretaceous, *4* Triassic and Jurassic, *5* Upper Cretaceous, *6* Eocene, *7* Oligocene and Miocene, *8* Pliocene, *9* Pleistocene

2.2 Tectonic of Sinai Peninsula

The region was formed as a result of tectonic plate movement: the African, Arabian and Mediterranean shields pulling away from each other, creating a great gash in the land. At the same time, the outward movement pushed against existing rock layers forming the looming Sinai Mountains. The effect was extensive, extending from what is now the Jordan River valley, south, then southeast, turning at Djibouti into Kenya, thus becoming known as the Great Rift Valley.

Two gulfs were created as branches to this valley. The Gulf of Suez leading to the Suez Canal, is relatively shallow (60–90 m) compared to the impressive depths of the Gulf of Aqaba. In this area dropoffs can disappear alarmingly, descents being between 1,000 and 2,000 m in some places.

The Peninsula of Sinai can be divided into a northern and southern region. The north consists of flat lying Paleozoic (550 million years) and more recent sediments, while the south consists essentially of metamorphic and magmatic rocks which are of Precambrian age (more than 600 million years). This southern portion is a continuation of the Arabo-Nubian Desert. A narrow belt (30 km wide) of soft Nubian sandstone also contains most of the Sinai minerals, e.g. turquoise, manganese.

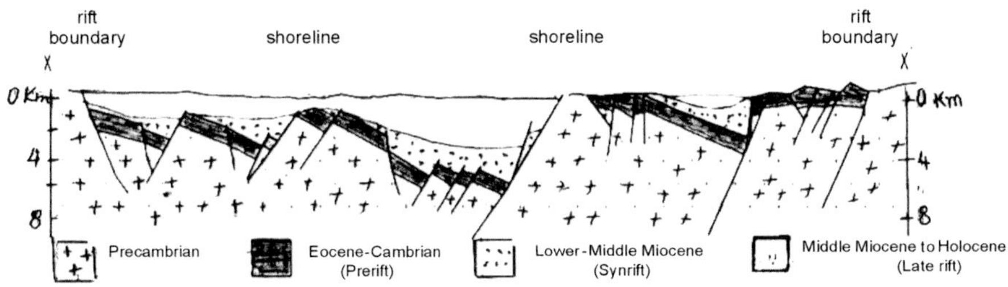

Fig. 4 Generalized schematic cross section across the Gulf and its coastal plains

The Gulf of Suez continental rift basin developed in the Upper Oligocene to Lower Miocene as the Red Sea opening propagated northward. The rift is structurally controlled largely by extensional normal faults that strike northwest, forming a complex array of tilted half grabens and asymmetric horsts (Pivnik et al. 2003).

The general stratigraphy and structural elements of the coastal plain are similar to, and continuous with, the structures and lithologies that define the Gulf of Suez rift. Within the rift and coastal plain domains, southwest-dipping normal faults and northeast-dipping bedding, characteristic of the central part of the rift, are observed (Figs. 4 and 5).

In contrast to the flat northern part, the granite mountains rise to peaks of more than 2,638 m in the south. The metamorphic rocks in Sinai were last formed about a thousand million years ago. At that time along the southern portion of Sinai, a north–south belt of volcanic islands were active following their deposition in a great oceanic basin.

This resistant magmatic rock now occupies some 60 % of the southern part of the peninsula and is responsible for the magnificent peaks (Gebel Serbal) and deepest canyons (Dahab). There were several periods of intrusion and dyke swarms before a brief burst of younger volcanic activity, Musa and Catherine.

The more pastel colored granites are much older than the reddish rock. These dykes (fine grained basaltic intrusions), having pushed their way into the original coarse grained plutonic granites, created stunning effects of color and form. To this is added the effects of sandstone weathering and coloring through leeched minerals. In some places, a whole hillside is layered in all colors, red, purple, green, cream, white, and orange. In many places, the evidence of past geological activity is obvious with tilted layers sitting on the underlying granite base rock.

3 Groundwater in Sinai Peninsula

The Nubian aquifer in the Sinai Peninsula extends laterally in the subsurface for approximately 10,000 km from the margins of the crystalline massif of central Sinai to the Gulf of Suez in the west and the Dead Sea in the northeast. The

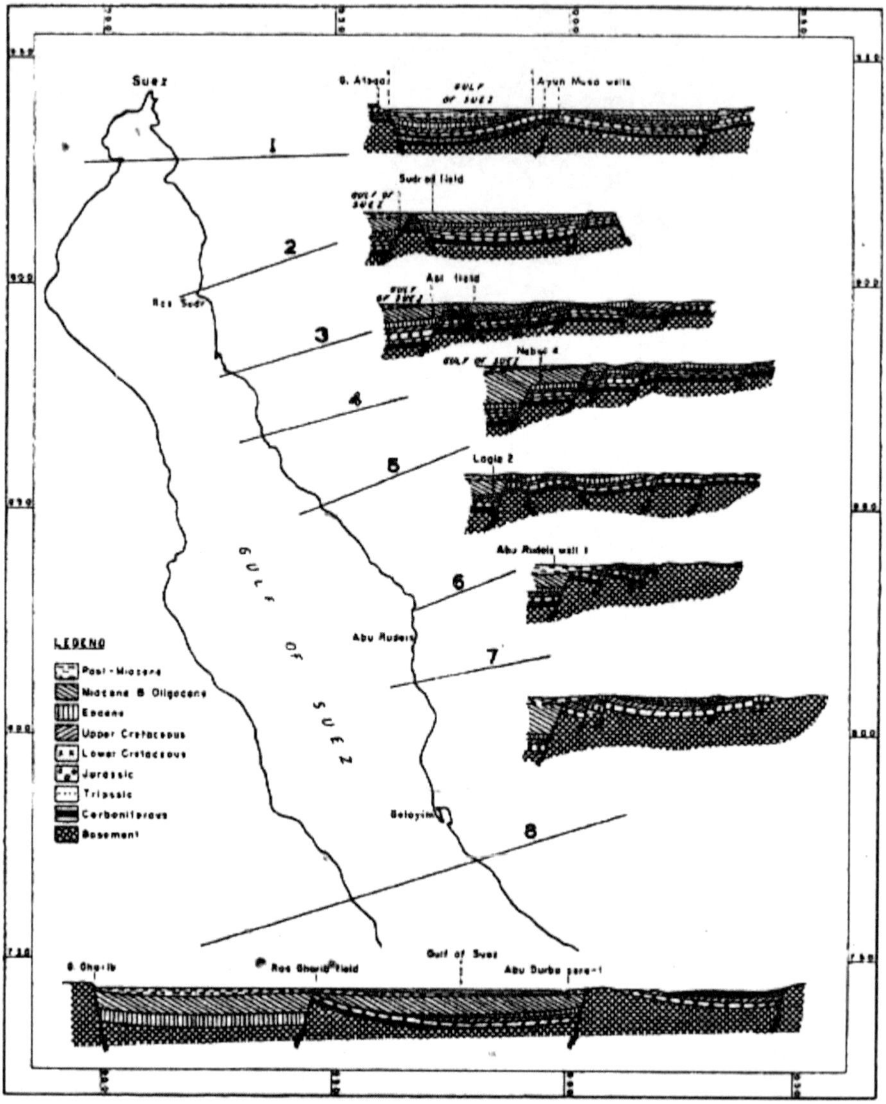

Fig. 5 Schematic geological cross-section through Gulf of Suez well fields (after Said 1962)

recharge is believed to have taken place through the sandstone outcrops at the northern flanks of the Precambrian basement complex in central Sinai and to a lesser extent through erosion windows in Sinai and Negev (Isaar et al. 1972). It has been suggested that in Pleistocene time, when more humid climatic conditions prevailed, that the Nubian sandstone aquifer was recharged by meteoric waters (Isaar et al. 1971, 1972; Isaar and Gilad 1982; Isaar 1987).

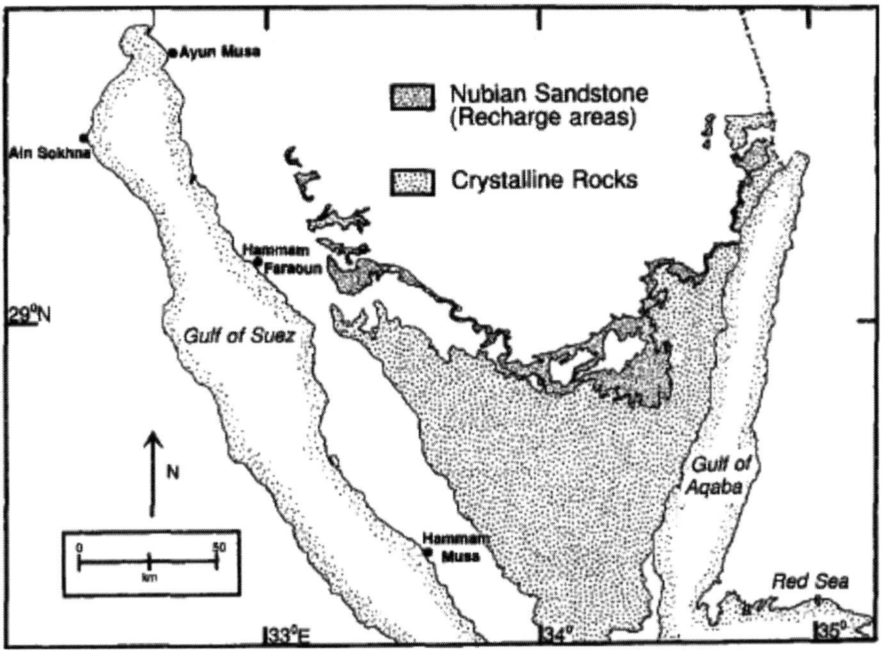

Fig. 6 Distribution of the crystalline basement, groundwater flow directions, and recharge areas of Nubian sandstone Aquifer in Sinai Peninsula (after Sturchio et al. 1996)

In Sinai, the precipitation over the Red Sea hills is channeled across the mountains where it reaches the Nubian sandstone outcrops at the foothills of the mountains. Some of the precipitation ends up infiltrating the Nubian sandstone outcrops and recharging the aquifer (Fig. 6). The groundwater flows downgradient towards discharge areas in the Mediterranean, Gulf of Suez, and cross the borders towards Israel (Isaar et al. 1972). The presence of major structural discontinuities along the Gulf of Suez allows Nubian Aquifer water at depth to access shallower environments. Some of the rising groundwater makes it to the surface and discharges in hot spring settings as is the case with Hammam Faraun, whereas other rising water might discharge in the thick alluvial sediments flooring the coastal plain as is the case with Hammam Musa before the water makes its journey to the final destination in the Red Sea.

The inflow of groundwater discharging along the Gulf of Suez is decreasing more than current precipitation, but not as much as Western Desert Nubian Aquifer groundwater. The Eastern Desert groundwater resembles the water of Sinai. It has intermediate composition between modern precipitation and the fossil water of the Western Desert as is the case with the Nubian aquifer groundwater samples in Sinai. It is believed that the Eastern Desert groundwaters are mixtures of fossil waters that were largely precipitated in previous wet climatic periods and distributions from modern precipitation that are precipitated in relatively drier periods. The time prevailing nowadays would be one of these dry periods.

4 Occurrence of Thermal Springs

Any water spring with a temperature exceeding 35 °C (95 °F) is considered as thermal. In Egypt, there are 19 recognized thermo-mineral springs.

Sinai possesses a wide range of rich physical features. Scientific research proved that clay found in some springs has therapeutic properties that can cure numerous bone, skin, kidney and gastrointestinal diseases and respiratory problems. The Red Sea water, together with its chemical composition and abundance of coral reefs, help cure psoriasis.

The following represents four important thermal water therapeutic sites in Sinai, namely Oyun Mousa, Ras Sidr, Hammam Pharaoun and Hammam Mousa, which are located on the eastern shore of the Gulf of Suez (Fig. 7). Opposite Oyun Mousa on the western bank of the Gulf of Suez is located the well-known Ain El Sukhna (hot spring in Arabic).

Geoelectrical survey at the localities having thermal surface manifestations could elucidate the groundwater aquifer, which is the main source of hot water in the vicinity of hot springs. At Oyun Musa, the water aquifer of Miocene is located around 300 m depth, which is very reasonable for exploitation. Meanwhile, at Hammam Faraun hot spring, the aquifer is a little deeper and thicker. The 2-D geoelectrical sections elucidate cross sections and give a reasonable explanation to the hypothesis of the origin of hot water sources at hot springs. The hypothesis states that, the hot water is deeply circulated and comes to the surface through the fault planes at the hot springs area (El-Qady et al. 2005).

The surface water temperature in these springs varies from 35 up to 72 °C. These springs owe their existence to tectonic heating associated with the opening of the Red Sea/Gulf of Suez rift (Boulus 1990).

4.1 Oyun Mousa

Oyun Musa (Moses springs) is a small oasis south of Suez. Twelve springs allow some large palms, tamarisks and other plants to flourish the land. It is said that this was the Hebrews' first camp after the exodus from Egypt. It is the very spot where, the legend goes, 12 water springs burst forth on Moses's visit.

Nubian Sandstone represents the main water aquifer at depth, which starts from 30 m. In this region the resistivity values measured are mainly governed by the lithological unit (El-Qady et al. 2005). The Miocene marl, sandstone and limestone are distinguished by varying resistivity attaining a value up to 150 Ohm/m.

Bentonite is found at Oyun Musa area and extends south to Sudr on the eastern coast of the Gulf of Suez; within the Miocene rocks it occurs in the form of thick beds in two zones separated by a marl bed. The upper bentonite bed is about 10 m in thickness while the lower bed is from 60 to 140 m, below a cover ranging in thickness from a few meters to 35 m. Faults caused some parts to be raised as blocks with little or no cover.

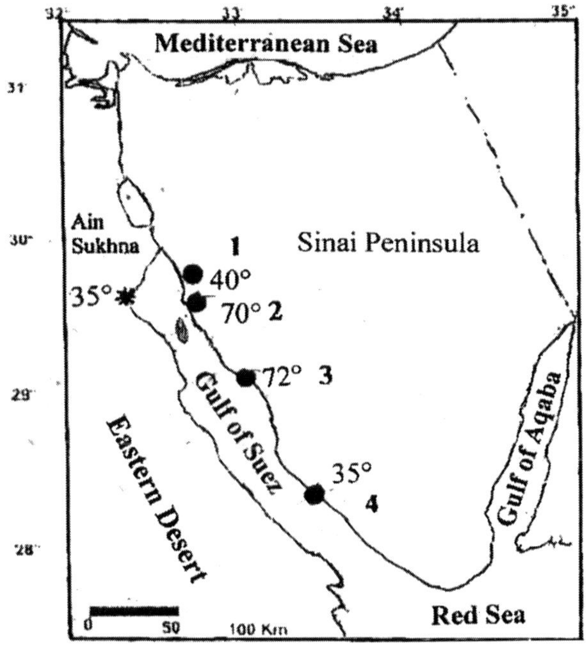

Fig. 7 Position of thermal springs and its surface water temperatures. *1* Oyon Musa, *2* Ras Sidr, *3* Hammam Faraoun, *4* Hammam Musa

4.2 Ras Sidr

Situated some 190 km from Cairo and 45 km from Ahmed Hamdi Tunnel, which connects mainland Egypt to Sinai, Ras Sidr is the second most important town on the Gulf of Suez after El Tor. Ain Musa is located about 20 km north of Ras Sidr. Ras Sidr is a floating berth oil-loading terminal situated on the Sinai Peninsula. The thermal spring is located among the oil-producing wells, which might be the reason why it is not often visited. The temperature of the lake water produced by the spring may be more than 70 °C.

4.3 Hammam Pharaoun Area

Approximately 45 km south of Ras Sidr (250 km from Cairo), there is the Pharaoh's Bath called Hammam Pharaon in Arabic.

The Pharoah's Bath Thermo Mineral spring with a year round temperature of 72 °C in Sinai is probably considered one of the most famous; and many visitors descend on the area each year to bathe in its mineral-rich mud and waters.

It consists of a group of natural hot sulfuric springs; the water flows directly from the mountain into a 100 m long natural lake right beside the seashore. A small cave has been carved out of the mountainside above for use as a natural

Table 1 Concentration of major ions (in ppm) in thermal and mineral waters od Sinai Peninsula

Parameter	Ain Helwan	Sudr, well 20	Hammam Musa	Ain El-Sukhna[a]	Hammam Faroun
TDS	4,045	7,885	5,082	7,532	1,6166
K^+	12.5	–	46.4	49.21	148.4
Na^+	1011.4	1,502	2486.8	1748.9	4443.6
Mg^{2+}	150.6	145	512.8	321.1	526.3
Ca^{2+}	303	691	310.6	418.8	644.3
Cl	1388.2	3440	3620.5	3546.4	7659.5
SO_4^-	1214.9	382	3081.7	988.9	2544.2
HCO_3^-	134.1	159	91.4	183.5	60.9
CO_3^-	156.1	0	43.8	0.0	108.1
PO_4^-	9	–	0.95	2.38	9.5
NO_3^-	3.7	–	6.2	18.6	1.8
Water conductivity [μS/cm]	8091	–	10164	15064	32332
Temperature [°C]	–	80 (surface)	37	35	35
Water type	Brackish	Saline, sulphuric	Saline, sulphuric	Saline, slight, sulphuric	Saline, slight, sulphuric

(–) indicates that the parameter was not precisely measured
[a] Hot spring located on the western side of the Gulf of Suez, opposite Ras Sidr

sauna. It has been proven that the water from the spring can help alleviate a number of diseases, including rheumatism, rheumatoid pains, kidney diseases, lung inflammation and several skin diseases. The water, smelling slightly unpleasantly of sulphur, bubbles from the rock inside a cave and flows down into the sea.

The subsurface structures are differentiated into two main faulting systems, in NNW-SSE and E-W directions, which explain the origin of the hot water that is deep circulating along the fault zone. Moreover, a huge thick aquifer about 100 m in the vicinity of the hot spring was elucidated. A promising area for geothermal drilling was around the Hammam Faraun; a hot spring was recommended and a project pumping this hot water for medical remedy is currently established.

4.4 Hammam Mousa Area

Hammam Musa (Moses' Spa) and a small oasis town is located 3 km west of the town of Al-Tur, the capital of southern Sinai. It is a wonderful oasis covered in palm trees around a natural hot spring, the water of which is capable of healing wounds especially of diabetic patients, treats the body weakness and regulates high blood pressure. The spring's water is saline and has a good deal of mineral concentration (Table 1; Fig. 8).

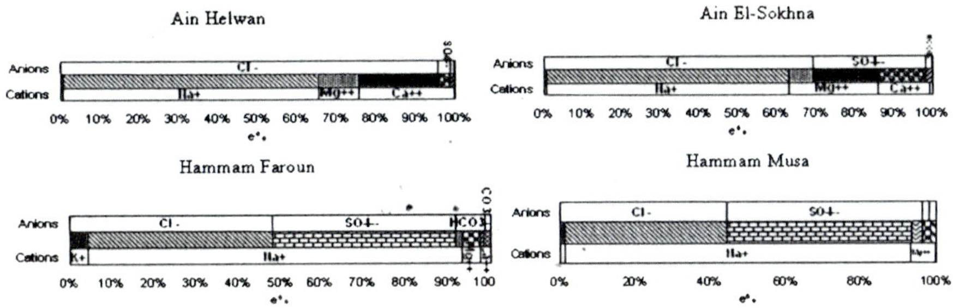

Fig. 8 Chemical composition of selected thermal waters of the Sinai Peninsula

Fig. 9 View on the Hammam Musa thermal springs

Moses' Spa consists of five natural sources. Many think that the water sources have medicinal properties. It has a large content of magnesium, sodium and sulfates. Hammam Musa is surrounded by palm trees and beautifully located in the wilderness, so that it looks like an oasis (Fig. 9).

4.5 Groundwater of Wadi Feiran

At present Wadi Feiran flows out through a narrow bottleneck cut into the Neogene marls and gypsum, to the sea north of the range of Qabeliat. Water wells drilled into this bottleneck, intended to exploit the underflow of the Wadi, encountered very shallow (20 m) alluvium and no exploitable water reserves. Detailed investigation revealed that the river underflow finds its way into Neogene sandstone and partly into the Nubian sandstones folded against the basement and the alluvium. This is possible owing to the fact that, until quite recently, Wadi Feiran did not flow directly to the sea, but drained mainly into the Qaa. This alluvial fan can be traced along the northern margins of El Qaa and overlies clastic Neogene rocks, in which brackish groundwater was found. Although crystalline rocks are believed to be impermeable, it was found that, owing to the tectonic history of the region, theses rocks form, owing to their fractures, a water-transmitting medium, which together with alluvial deposits and down faulted blocks, may transfer between 10 and 15 % of the total precipitation in the region. In some parts of the area (as in the lacustrine deposits of the main wadis, in the sedimentary fill of the rifts and in the sedimentary basins bordering the crystalline province), large volumes of water suitable for exploitation during drought years may be located (Isaar and Gilad 1982).

References

Boulus F (1990) Some aspects of the geophysical regime of Egypt in relation to heat flow, groundwater and micro earthquake. In: Said R (ed) The Geology of Egypt. A. A. Balkema, Rotterdam

El-Qady G, Salem A, Aboud E, Khalil, Ushjima K (2005) Geothermal reconnaissance study for Sinai Peninsula, Egypt. In: Proceedings of World Geothermal Congress, Antalya, Turkey

Isaar A (1987) Fossil water under the Sinai-Negev Peninsula. Sci Am 253(1):104–110

Isaar A, Bein A, Michaeli M (1972) On the ancient water of the Nubian sandstone aquifer in Central Sinai and Southern Israel. J Hydrol 17:353–374

Isaar A, Gilad D (1982) Groundwater flow systems in the arid crystalline province of Southern Sinai. Hydrol Sci 27(3):309–325

Isaar A, Rosenthal E, Eckstein Y, Bogoch R (1971) Formation water, hot springs and mineralization phenomena along the eastern shore of the Gulf of Suez. Bull Int Assoc Sci Hydrol XVI(3):25–44

Kora M (1995) An introduction to the stratigraphy of Egypt. Lecture Notes. Geology Department, Michigan University, 110 p

Pivnik DA, Ramzy M, Steer BL, Thorseth J, El Sisi Z, Gaafar I, Garing JD, Tucker RS (2003) Episodic growth of normal faults as recorded by syntectonic sediments, July oil field, Suez rift, Egypt. AAPG Bull 87:1015–1030

Reynolds ML (1979) Geology of the Northern Gulf of Suez. Ann Geol Soc Egypt v(IX):322–343

Said R (1962) The geology of Egypt. Elsevier, Holland 377 p

Said R (ed) (1990) The geology of Egypt. A. A. Balkema, Rotterdam, 734 p

Schütz KI (1994) Structure and stratigraphy of the Gulf of Suez, Egypt. In: Landon SM (ed) Interior rift basins, vol 59. AAPG Memoir, pp 57–96

Sturchio NC, Sultan M, Sanyo Y, Abo Kamar Y, Sayed M (1996) Composition and origin of thermal waters in the Gulf of Suez area, Egypt. Appl Geochem 11:471–479

Enhanced Fluoride in Groundwater in Eastern Anatolia: Effects, Origin and Possibilities for Remediation

Werner Balderer, Fanny Leuenberger, Giorgio Menghini and Walter Dierauer

Abstract In the spring of 2002 and summer of 2003 two research surveys were carried out in eastern Anatolia. The hydrogeological/hydrochemical investigations undertaken in the spring of 2002 revealed a dramatic situation of the water supply in several villages in the Dogubeyazit area, manifesting fluoride concentrations largely above the 1 ppm WHO limit of fluoride concentration in drinking water. The severity of the resulting health problems was confirmed by a dental and general health survey in three villages where the available water supply contained 6–8 ppm of fluoride. Consequently, this hydrogeological study will attempt to contribute to the alleviation of this problem by the following: (i) to survey the existing situation with regards to the current use of groundwater provided by the existing springs, and boreholes, (ii) to propose a new distribution network of groundwater resources with low fluoride concentration in order to lower its intake thus improving health and quality of life of the affected population.

Keywords Hydrogeological/hydrochemical investigations · Fluoride · Drinking water · Anatolia · Turkey

W. Balderer (✉)
ETH Zurich, Swiss Geotechnical Commission, SGTK, Sonneggstr. 5, 8092 Zurich, Switzerland
e-mail: werner.balderer@erdw.ethz.ch

F. Leuenberger
ETH Zurich, Geological Institute, Engineering Geology, Sonneggstr. 5, 8092 Zurich, Switzerland

G. Menghini
Division of Applied Prevention, Department of Preventive Dentistry, Periodontology and Cariology, Dental Institute of the University of Zurich, Zurich, Switzerland

W. Dierauer
Palu 7, 7023 Haldenstein, Switzerland

W. Balderer et al. (eds.), *Thermal and Mineral Waters*, Environmental Earth Sciences, DOI: 10.1007/978-3-642-28824-1_4, © Springer-Verlag Berlin Heidelberg 2014

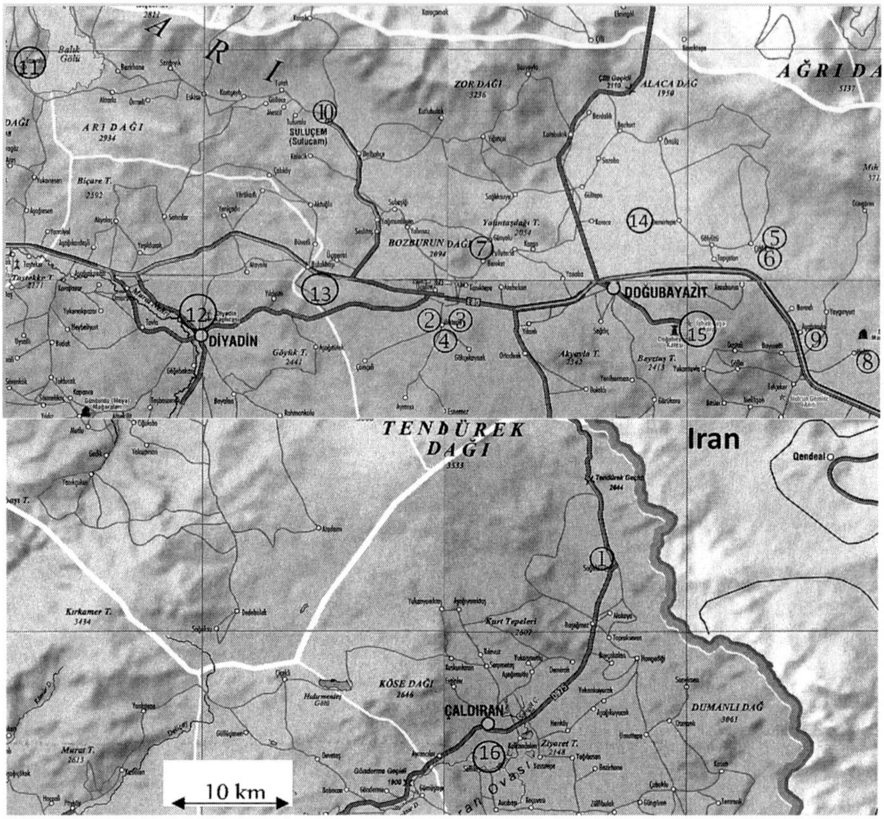

Fig. 1 Area of investigation in Eastern Anatolia, Turkey (numbers correspond to analyzed groundwater samples)

1 Introduction

In the spring of 2002 and the summer of 2003 two research surveys were carried out in eastern Anatolia. These were intended to evaluate the situation in the area regarding dental fluorosis and the hydrogeological conditions of groundwater resources, supposed to be responsible for the enhanced fluoride in the groundwater. The investigated area (Fig. 1) is situated north of Lake Van and situated between the main towns of Caldiran in the South, Dogubayazit in the central part and Igdir in the North, up to the Iranian border in the East and the Armenian-Azerbaijan border in the North. Between the mountains of volcanic origin as Agri

Fig. 2 Effects of fluorosis: 35 year old man

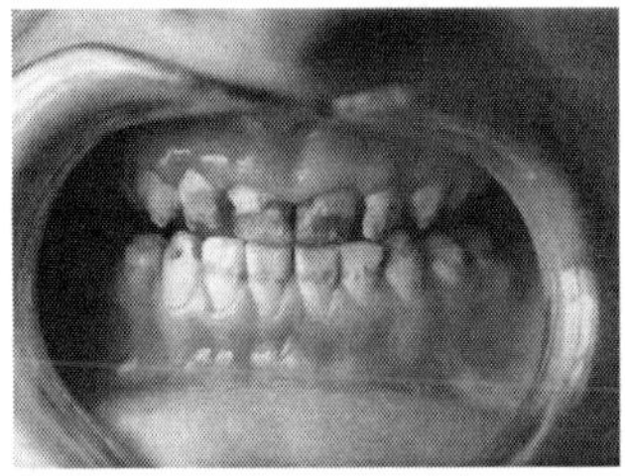

Fig. 3 Effects of fluorosis: child's teeth

(Ararat) and Tendürek extends a great plain which contains loose volcanic sediments and also alluvial deposits (Yılmaz et al. 1998).

1.1 Results of Dental Survey

An initial dental survey including the collection of urine spot-sampling and nail-sampling in three villages where the available water supply contained 6–11 ppm fluoride was carried out in 2003. From the investigation of 119 children aged between 4 and 15 years the following mean concentrations of fluoride resulted: 9 ppm fluoride in urine and 19 ppm F in the nails. The resulting severe dental fluorosis is illustrated by the photographs taken of men (Fig. 2) and children (Fig. 3) living in this area. The possible repercussions on the skeleton of the high fluoride concentration in drinking water were not considered in the initial preliminary investigations. However, over dozens of years, the regular intake of fluoride in the concentrations measured will undoubtedly lead to endemic skeletal fluorosis.

2 Main Question: Where does this Fluoride Rich Water Come from?

The hydrochemical investigations undertaken in the spring of 2002 revealed a dramatic situation of the water supply in several villages in the Dogubeyazit area, manifesting fluoride concentrations well above the WHO limit of 1 ppm in drinking water. The water which is used both as drinking water and as a general water supply originates almost exclusively from groundwater coming from natural springs, hand dug wells and drilled boreholes. Consequently, the source of fluoride is coupled with the origin of the groundwater and the chemistry of the rock/mineral environment.

3 Chemical Characterization According to the Results of the Investigated Groundwater

The main chemical composition of the investigated groundwater samples is represented in Table 1. In order to visualize these results the concentrations of the different ions are represented in two triangular (Piper-type) diagrams (Figs. 4, 5) as equivalent ratios and in Fig. 6 as equivalent concentrations in a Schoeller type diagram. These diagrams show the relationship between the measured dissolved major ion constituents of the investigated groundwater. According to the triangular diagrams most sampled groundwater are of bicarbonate type, Nr. 2 and 9 also containing higher concentrations of chloride and sulphate. Concerning the main cation composition a great variety of waters of Mg–Ca to Mg–Ca–Na and even to Mg–Na type exists.

From the Schoeller type diagram of Fig. 6 the great disparity between the concentrations of the dissolved main ionic components is also apparent. The highest total concentration of dissolved solids is found in the thermal water of Nr. 12, Aydin Kaplica (Kaplica meaning spa in Turkish).

The two parameter diagram of fluoride concentration versus electrical conductivity (Fig. 7) denotes a slight increase in fluoride concentration with increasing total mineralization as shown by a rise in electrical conductivity ($r^2 = 0.43$). However, a correlation between the bicarbonate concentration, as stipulated by Özgür et al. (1992a, b) and Oruç et al. (1976) for investigated waters of the Tendürek area (with r^2 equal to 0.09 and 0.26, respectively), could not be confirmed by these analyses.

4 Interpretation

Following the analyses of different samples of drinking water taken from drilled boreholes and dug wells, as well as captured spring outflows the origin of the enhanced fluoride may be postulated to involve the following processes:

- Atmosphere: Atmospheric particles (originating from natural dust as wind deposits of the large arid plains and/or air pollution) may be dissolved in the atmospheric humidity (moisture or even rain) and reach the earth surface as rain or snow.

Table 1 Results of chemical analyses of the investigated waters (by Ion-Chromatography)

Sample location	Nr.	Ion concentrations (mg/l)											TDS (mg/l)
		Li^+	Na^+	K^+	Mg^{2+}	Ca^{2+}	Sr^{2+}	F^-	Cl^-	NO_3^-	SO_4^{2-}	HCO_3^-	
Soguk Su Spring	1	0.1	51.3	6.7	29.9	58.1	–	1.1	6.9	7.7	6.0	500.2	668.0
Alintepe, hp.	2	0.1	356.9	20.4	77.1	51.9	2.1	13.5	212.6	145.0	272.7	744.2	1,896.5
Alintepe, Spring	3	0.1	165.4	11.9	4.6	10.2	–	11.1	10.8	5.0	14.8	457.5	691.4
Alintepe good sp.	4	0.1	1.6	4.2	13.7	47.4	0.8	0.6	7.9	30.4	21.5	207.4	335.6
Ciftlik-Cekmet, sp.	5	0.1	190.7	10.5	39.5	41.1	–	8.0	41.9	12.8	32.6	793.0	1,170.2
Cevekmet,art. bh.	6	0.1	419.4	15.6	45.8	23.6	–	14.9	49.1	12.8	56.0	1,238.3	1,875.6
Pullutarla bh.	7	0.1	190.2	9.3	37.2	41.9	–	8.4	41.5	11.1	32.1	762.5	1,134.3
Halaç bh.	8	0.1	196.7	11.4	30.7	26.5	1.6	6.8	52.5	28.9	96.4	610.0	1,061.5
Asagitavla Köy bh.	9	0.1	142.7	3.5	30.3	76.2	3.3	13.3	55.8	44.6	247.2	341.6	958.6
Suluçam Köyu sp.	10	0.1	4.1	2.4	3.2	7.5	–	0.7	3.4	4.4	16.9	61.0	103.7
Balik Lake water	11	0.1	6.4	1.9	5.0	20.5	–	0.8	4.1	1.1	4.7	103.7	148.3
Aiydin Kaplica	12	0.1	160.1	47.9	78.3	453.0	4.6	9.5	136.4	n.a.	127.6	1,982.5	3,000.0
Dogubayasit Plain	13	0.1	140.7	12.2	112.2	59.9	n.a.	114.9	296.2	n.a.	n.a.	n.m.	736.2
Cemal (lake)	14	0.1	348.6	12.2	14.3	11.9	–	8.9	52.4	n.a.	61.9	1,866.6	2,376.9
Village above Dogubayasit	15	0.1	31.8	2.8	11.4	28.9	0.5	1.4	6.1	10.9	12.5	274.5	380.9
Selik village sp. n. bridge	16	0.1	120.7	14.0	84.8	183.0	–	6.2	36.9	n.a.	17.3	1,372.5	1,835.5

Note n.a. not available, no traces found, *n.m.* not measured, HCO_3^- concentrations determined by titration, *TDS* sum of analyzed constituents

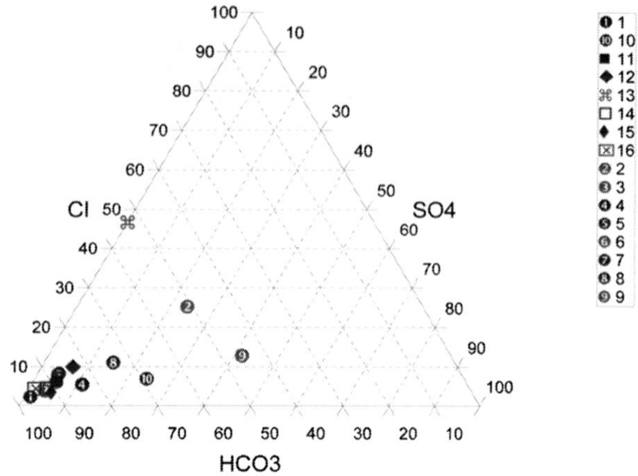

Fig. 4 Triangular (piper type) diagram of cation composition of groundwater (meq-% ratio). *Numbers* correspond to Fig. 1 and Table 1

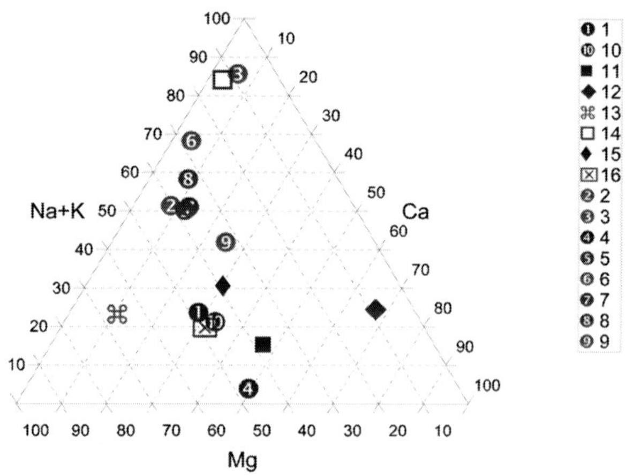

Fig. 5 Triangular (Piper type) diagram of anion composition of investigated groundwater (meq-% ratio). *Numbers* correspond to Fig. 1 and Table 1

- Rock environment along the groundwater flow path (aquifer): In the investigated area, deposits of volcanic activity within the quaternary geological time (such as ash deposits, lava flows, etc.) are present as an effusion of today's quiet volcanoes of Tendürek and Agri. The basic composition of the groundwater may be changed through chemical interaction of the groundwater with the minerals contained in the aquifer rocks. Depending on the origin of the rocks (of magmatic, metamorphic or sedimentary origin) a vast variety of minerals may be present (Senior and Sloto 2006; Stormer and Carmichael 1970).
- Deep underground sources (upflow of deep seated fluids originating from formation fluids or from magma) along tectonic fault zones. Such NE–SW trend fault zones were identified in the area of Caldiran between Dogubayazit and the Balik Lake.

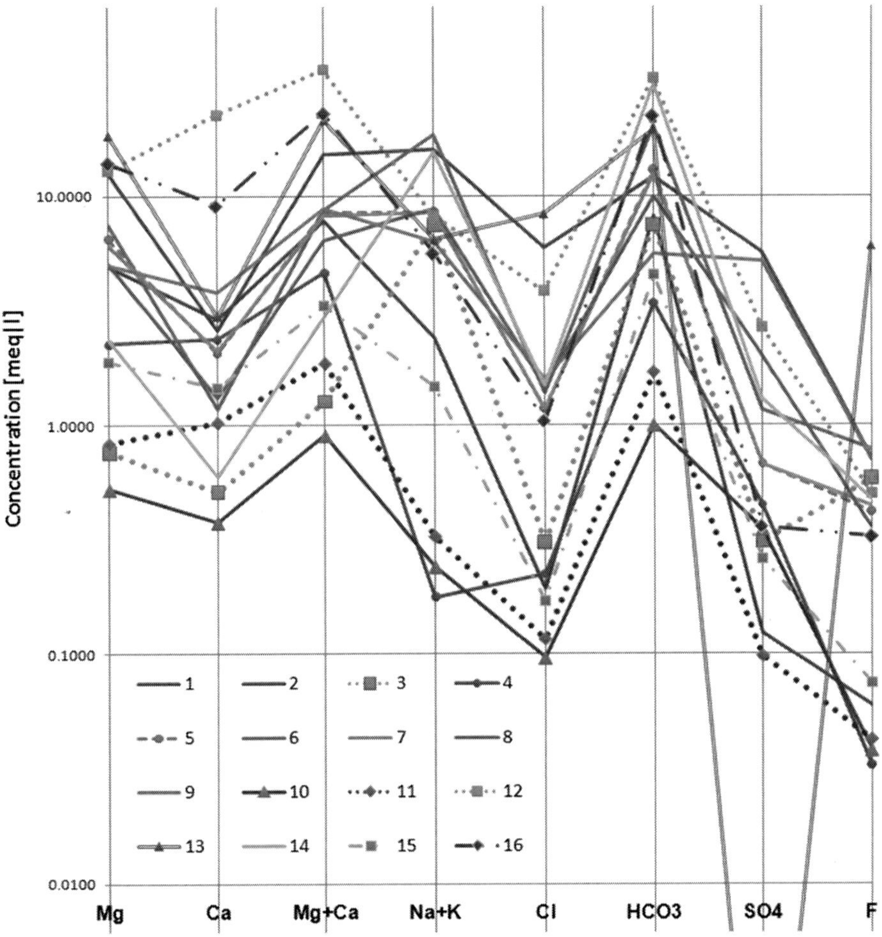

Fig. 6 Characterization of the main chemical composition of the investigated groundwater (Schoeller-type diagram). *Numbers* correspond to Fig. 1 and Table 1

- Within such fault zones heavily mineralized fluids are often upwelling and mixed with the normal groundwater of shallow aquifers. These fluids of deep origin are heavily mineralized, (e.g. formation brines) and present very long residence times depending on their origin either as evolved original sedimentary pore fluids or as condensed fluids of magmatic steam and gases.

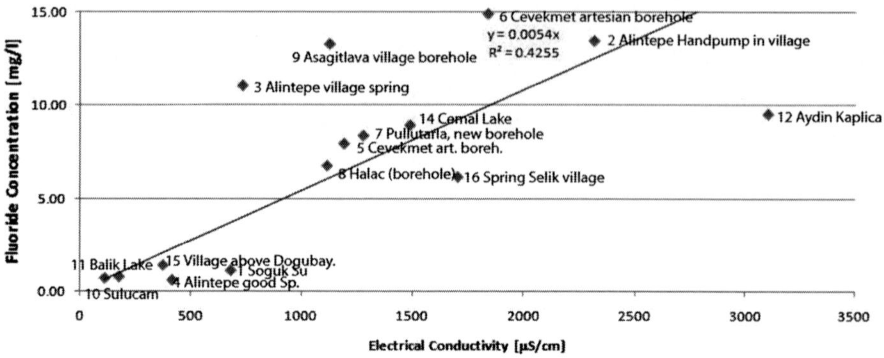

Fig. 7 Fluoride concentration versus electrical conductivity (mS/cm) of the investigated groundwater. *Numbers* correspond to Fig. 1 and Table 1

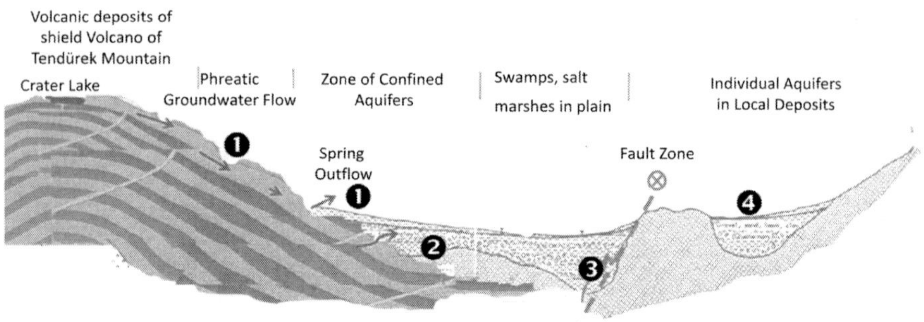

Fig. 8 Schematic cross section of the hydrogeological situation of the Dogubayazit area (modified after Wallick and Toth 1976)

5 Conclusions Concerning Origin of Fluoride Rich Waters Based on Hydrogeological Observations

From observations at the origins of springs as well as from the situation of boreholes the following characteristics are seen. These are also summarized in the conceptual model in Fig. 8:

1. Water of shallow origin with short residence time and flowing with subsurface flow systems contain essentially low mineralization as well as low fluoride concentration. This also applies to fast flowing waters in rivers and lakes having a constant outflow. These characteristics mainly apply to non-confined groundwater.
2. Water with somewhat longer residence time along a longer flowpath is characteristic for groundwater within the non-consolidated sediments in the large alluvial plains under confined or artesian conditions. The chemical equilibrium

of these waters are complete and contain a higher amount of fluoride as was observed in dug wells and drilled boreholes (into the alluvial infilling, usually at depths of 50 to about 200 m below ground surface).

3. Groundwater in aquifers which are transected by fault zones: In such a situation the groundwater is affected by upflowing deep water with higher mineralization and temperature or higher gas content (mainly CO_2), which causes an upwelling (buoyancy) effect. The same situation also applies for lakes situated on active tectonic fault zones.
4. Shallow water in greater plains, in final depression sites, where more water evaporates than flows through this swamp area (with only periodic outflows during the spring season), accumulates salt by evaporation. Consequently, very high fluoride concentration is revealed, as shown by the data of the swamp sample 13 which contains 115 mg/l of fluoride.

6 Possibilities of Remediation of the Existing Drinking Water Supply Based on the Observed Hydrogeological Conditions

Possible subsequent improvements based on the observed hydrogeological conditions remediation without using an extensive fluoride removal system can only be achieved by collective solutions.

This would imply the capture of water of low fluoride concentration of young shallow groundwater and its further distribution to the villages of the whole area. Such groundwater exists as proved by the samples of Nr. 1, 4, 10, 11, 15. Furthermore, another possibility would be a dilution of the fluoride content of existing groundwater resources by inputs of rainwater and river water. For this purpose the rain and river water should be collected and stored in appropriate reservoirs.

Such solutions would, on a social political level, lower the power of the local authorities and compel them to search for more communal solutions. Unfortunately, this is in direct opposition to the independency concern of the local village ruling system. Social reforms would be needed; however, presently it is very difficult to establish due to the unstable political situation of the region. Therefore, by drilling a borehole in each village as a source of a new drinking water supply would not alleviate the problem but rather accentuate it as shown by the results of analyzed fluoride concentrations of the water from newly drilled boreholes as Nr. 5, 6, 7 and 9.

What is needed is a local to regional development scheme which is implemented by the local population itself.

References

Oruç N, Alpman N, Karamanderesi H (1976) Hydrogeology of the spring waters with high F content from the surroundings of Tendürek volcano. Bull Geol Soc Turk 19:1–8
Özgür N, Pekdeğer A, Schneider H-J (1992a) High fluorine contents in the Pliocene volcanic rocks of the Gölcük area, Isparta, SW Turkey. Bull Geol Soc Greece XXVIII/2:417–427
Özgür N, Pekdeğer A, Schneider H-J (1992b) Fluorine in Pliocene volcanic rocks of the Gölcük area, SW Turkey. In: Proceedings of the 1st international symposium on East Mediterranean geology, Adana, Turkey
Senior LA, Sloto RA (2006) Arsenic, boron, and fluoride concentrations in ground water in and near diabase intrusions, Newark Basin, Southeastern Pennsylvania. Agency Scientific Investigations Report 2006-5261, U.S. Department of the Interior U.S. Geological Survey
Stormer JC Jr, Carmichael ISE (1970) Villiaumite and the occurrence of fluoride minerals in igneous rocks. Am Min 55:126–134
Wallick EI, Toth J (1976) Methods of regional groundwater flow analysis with suggestions for the use of environmental isotope and hydrochemical data in groundwater hydrology. IAEA, Vienna, pp 37–64
Yılmaz Y, Güner Y, Saroglu F (1998) Geology of the quaternary volcanic centers of the east Anatolia. J Volcanol Geoth Res 85:173–210

Origin of Thermal Waters in Budapest Based on Chemical and Isotope Investigations Including Chlorine-36

Werner Balderer, H. Arno Synal, J. Deák, I. Fórizs and Fanny Leuenberger

Abstract Budapest is a major spa center with numerous thermal baths that are open to the public. Thermal spas in Budapest were first developed by the Romans and followed by the Turks, present spas were built mainly in the 19th and 20th centuries. At the city of Budapest the Danube River flows along a geological fault which separates the Buda Hills from the Great Plain. Within this fault zone in the vicinity of the Danube more than 100 thermal springs are arising yielding totally about 40,000 m^3 per day of warm mineral water. In this study the results of thermal, chemical and isotope analyses (including tritium and chlorine-36) of 12 thermal springs and wells are presented. These results are interpreted with respect to the origin and recharge conditions of the investigated thermal waters.

Keywords Thermal waters · Hydrogeological processes · Isotopes · Budapest

---

W. Balderer (✉)
Swiss Geotechnical Commission, ETH Zurich, SGTK Sonneggstr. 5,
8092 Zurich, Switzerland
e-mail: werner.balderer@erdw.ethz.ch

H. A. Synal
Institute of Particle Physics, ETH Zurich, Schafmatt Str. 20,
8093 Zurich, Switzerland

J. Deák
GWIS Ltd, Alkotmány Street 45, Dunakeszi 2120, Hungary

I. Fórizs
Hungarian Academy of Sciences, Institute for Geochemical Research,
Budapest, Budaörsi út 45, Budapest 1112, Hungary

F. Leuenberger
Geological Institute, Engineering Geology, ETH Zurich, Sonneggstr. 5,
8092 Zurich, Switzerland

W. Balderer et al. (eds.), *Thermal and Mineral Waters*, Environmental Earth Sciences,
DOI: 10.1007/978-3-642-28824-1_5,

1 Introduction

The capital of Hungary is situated along the Danube River, in the heart of the Carpathian basin. Budapest is divided by the Danube in two parts, the higher hilly side of Buda and the lower plain named Pest on the other side; therefore, both sides together form Budapest. Hilly Buda, which comprises one-third of the city's area of 525 km^2, is located along the right bank of the Danube surrounded by low mountains up to an altitude of 529 m above msl.

Budapest is a world known famous spa center with numerous thermal baths (Fig. 1).

The origin of these naturally outflowing thermal waters is due to the special geological features of this area in central Hungary. A conceptual hydrogeological model is presented in Fig. 2.

At the city of Budapest, the Danube River flows along a geological fault of north–south direction, which separates the Buda Hills from the Great Plain (Fig. 2). Within this fault zone in the vicinity of the Danube more than 100 thermal springs are arising, yielding a total of about 40,000 m^3 per day of warm mineral water.

As shown on the geological cross section of Fig. 3, the geological situation can be summarized as follows: the Danube (thermal) fault zone divides the city of Budapest in two distinct areas. In the Buda Hill at the Western bank of the Danube, Late Triassic dolomites are exposed forming a horst-anticline structure. On the other side of the Danube River the Pest plain extends to the east. In this area the Triassic dolomites and Eocene limestone are covered by several 100 m thick Paleogene shelf margin deposits and Quaternary alluvium of the Danube. In previous publications (Alföldi et al. 1978; Korpás et al. 1999), the mechanism of the ascending thermal waters is explained by the following theory: the karst water circulation within these dolomitic limestone formations is controlled partly by depositional features (unconformities, caves and clastic horizons) and partly by tectonic features such as faults and related fractures. "Ascending thermal water (along faults) mixes with cooler meteoric water giving the thermal springs and spas along the Danube" (Alföldi et al. 1978; Korpás et al. 1999).

1.1 Aim of the Present Study

The hydrogeological situation of the thermal springs is explained by the above-mentioned mixing theory; however, it remains still to prove these postulated hydrological mixing processes by investigation of the properties of the outflowing groundwater itself.

As demonstrated in Balderer et al. (2004), for the infiltration of the Danube River in western Hungary (Szigetköz area) the two radionuclide tritium (^{3}H) and Chlorine-36 (^{36}Cl) are well suited to study the infiltration process of recent

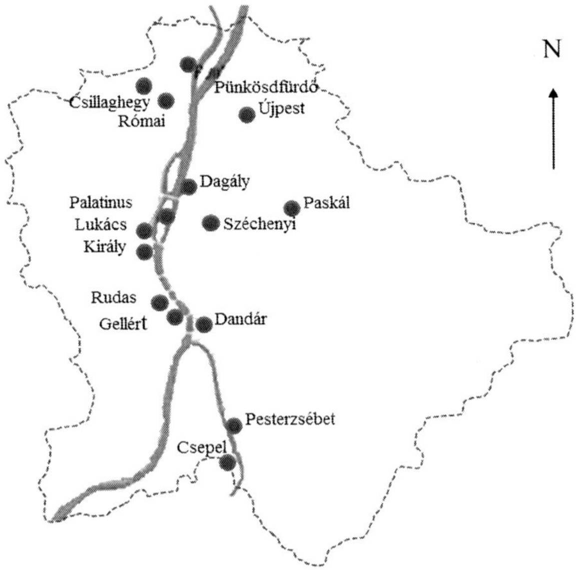

Fig. 1 Map with the Spa area of Budapest

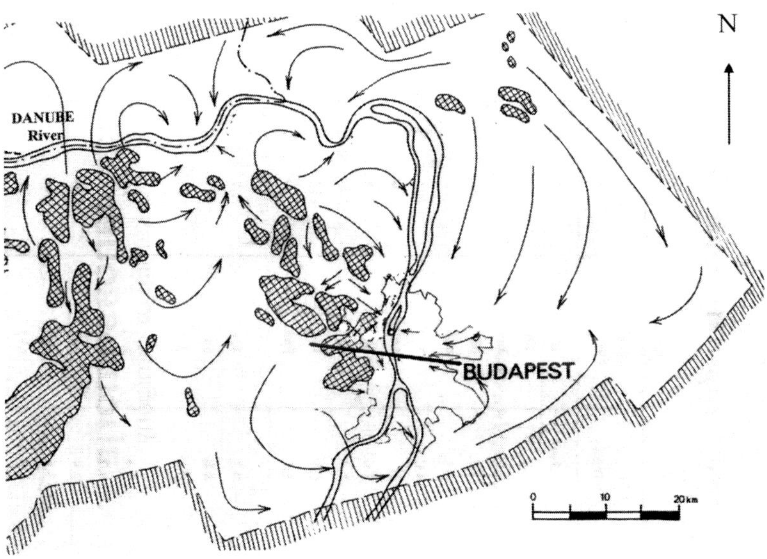

Fig. 2 Conceptual model of Budapest thermal karst-water regime after Alföldi et al. (1978) and Deák (1979). The bold straight line presents the orientation of the cross-section presented in Fig. 3. Karstified rocks are on the surface on the grated areas (recharge). Arrows show the possible flow paths of karst-water

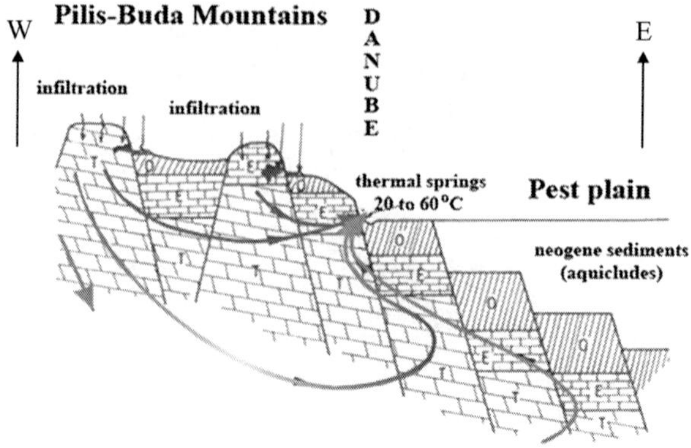

Fig. 3 Hydrogeological cross section presenting the water circulation in the Budapest thermal karst regime (after Kovács and Müller 1980). Arrows indicate schematic flow paths which are 3-dimensional and step out from the cross section in both the south and north direction (See Fig. 2). Used Terms of lithological units: *T* Triassic dolomite; *E* Eocene limestone; *O* Oligocene clay

groundwater along a groundwater flow path as still some bomb-produced nuclides are remaining within the water cycle.

In the present situation of mixing recent recharged karst groundwater with postulated groundwater of deep origin at higher temperatures and also of different chemical composition, the two mentioned radionuclides provide additional properties which are well suited to check for proposed hydrogeological processes of the formation of the Budapest thermal waters.

In a previous investigation of the groundwaters of the quaternary aquifer of the Great Hungarian Plain, the ^{36}Cl method proved to be applicable also for groundwater of recharge within the last glacial period; and it was also in agreement with the results of the ^{14}C and the stable isotopes ^{18}O and ^{2}H (Balderer et al. 1998).

2 Introduction to the Application of ^{36}Cl in Groundwater Studies

2.1 Origin of Chlorine-36 in Groundwater

^{36}Cl with a half-life of about 301,000 years is one of the most promising isotopes for environmental studies. The ^{36}Cl content is usually expressed as concentration with respect to the volume of water (number N of ^{36}Cl atoms per litre of H_2O) or as the atomic ratio R of ^{36}Cl with respect to the total chloride.

$$\text{The conversion factors are: } N = 1.7 \cdot 10^{19} \cdot R \cdot C \quad (1)$$

$$\text{or: } R \cdot C = N \cdot 5.89 \cdot 10^{-20} \quad (2)$$

where, C is the chloride content of the water in mg/l. The principles of the ^{36}Cl method are well described by several authors as reviewed in Balderer and Synal (1996, 1997).

The main cycle in groundwater is related to the following processes and reservoirs.

2.1.1 Atmospheric Input

Natural ^{36}Cl is primarily produced in the stratosphere in spallation reactions of secondary cosmic protons and neutrons on ^{40}Ar. Its atmospheric residence time is about 1–2 years. From the atmosphere it is removed and subsequently stored in terrestrial deposits. Due to atmospheric mixing processes the ^{36}Cl input to these deposits is expected to show latitude dependence (Bentley et al. 1986). New ^{36}Cl measurements on ice cores from Greenland at latitudes between 60° and 70°N result ^{36}Cl fallout rates of about 20 atoms/(m^{-2}/s) (for the period not affected by the anthropogenic thermo nuclear production; Synal et al. 1990).

This is approximately a factor of 2 higher than the expected fallout rates according to Bentley et al. (1986).

The ^{36}Cl concentration N_o for the infiltration water is given by the following equation:

$$No = F/P \cdot 3.156 \cdot 10^{7} \cdot 100/(100 - E) \quad (3)$$

with:

F Fallout rate of ^{36}Cl (at m^{-2} s^{-1})
P Mean annual precipitation of the considered catchment area (mm)
E Mean annual evapotranspiration rate (in % of P).

2.1.2 Anthropogenic Thermo Nuclear Production

Nuclear weapon tests in the South Pacific in the years 1952–1959 produced large amounts of ^{36}Cl through the reaction:

^{35}Cl (n,γ) ^{36}Cl by neutron activation of chloride contained in sea water. ^{36}Cl contents, which are about three orders of magnitude above the normal level in precipitation waters, have been observed in ice cores from Dye-3 in Greenland between 1952 and 1956 (Synal et al. 1990).

In shallow groundwater also anthropogenic produced tritium and increased ^{36}Cl concentration were measured as reported in Bentley et al. (1986).

Table 1 $^{36}Cl/Cl$—ratios of achieved secular equilibrium of underground production

Rock type	$^{36}Cl/Cl$—ratio
Vulcanite	$10\ (\pm 1.7) \cdot 10^{-15}$
Rhyodacite	$19\ (\pm 1.1) \cdot 10^{-15}$
Marble (limestone)	$5.6\ (\pm 0.4) \cdot 10^{-15}$
Travertine	$0.9\ (\pm 0.3) \cdot 10^{-15}$

2.1.3 Near Surface Production

In case of calcium carbonate formations, ^{36}Cl production from cosmic myons is of importance down to depths of 100 m below surface.

2.1.4 Radiogenic Production in the Deep Subsurface

Within the geological environment ^{36}Cl is naturally produced by the radiogenic reaction $^{35}Cl\ (n,\gamma)\ ^{36}Cl$ and expressed by the following equation:

$$R_{eq} \cdot C_o \cdot (1 - e - \lambda t) \qquad (4)$$

where C_o is the chlorine content of the recharge groundwater and R_{eq} the $^{36}Cl/Cl$ ratio for achieved equilibrium of underground production (depending on the radio element content of the aquifer rock). An additional source presents dissolved chloride originating from the aquifer rocks which is contained in minerals, fluid inclusions and as inter granular salt in pore spaces. As this chloride is present in the rock for geological times it can be assumed that the ^{36}Cl has already achieved the secular equilibrium of production within the surrounding rock environment.

In order to give the magnitude of thc possible range of underground production for a groundwater of long residence time, the following $^{36}Cl/Cl$ ratios of achieved secular equilibrium of underground production (that means a residence time of more than 5 half-lives, that means more than 1.5 million years) are given in Table 1, according to Balderer and Synal (1997).

2.1.5 Radioactive Decay During Residence in the Aquifer Along Flow Path

Within the groundwater the atmospheric component is subject to radioactive decay.

The corresponding equations are:

$$\text{if expressed as ratios: } R = R_o \cdot e^{-\lambda t} \qquad (5)$$

and,

$$\text{if expressed as concentrations: } N = N_o \cdot e^{-\lambda t} \tag{6}$$

$$\text{according to the equation: } RC = R_o \cdot C_o \cdot e^{-\lambda t} \tag{7}$$

2.1.6 Residence Time of the ^{36}Cl in Groundwater

The resulting residence time of the ^{36}Cl in groundwater is determined by taking into account all these processes by the following equation derived from mass balance a mass balance approach:

$$t = -1/\lambda \cdot \ln\{C/C_o \cdot (R - R_{eq})/(R_o - R_{eq})\} \tag{8}$$

where:

λ ln2/$T_{1/2}$ = decay constant of ^{36}Cl,
$T_{1/2}$ half-life of ^{36}Cl equal to 301,000 years
R ^{36}Cl/Cl ratio of water sample (R),
R_o ^{36}Cl/Cl ratio of infiltrating water
R_{eq} ^{36}Cl/Cl ratio of chloride in secular equilibrium of production
C_o chloride concentration of the infiltration water (directly from precipitations)
C chloride concentration of the water sample.

3 Interpretation of the Results of Analyses of the Sampled Waters

According to the obtained data of the chemical and isotope analyses as presented in Table 2 the following characteristics can be deduced:

- The water of the cold spring of some km NW to Budapest on the hilly recharge area (Csobánka, Szentkút OBP21) yields the lowest water temperature of 10 °C, the highest tritium concentration of 4.3 TU, and the highest ^{36}Cl/Cl ratio of 1007 $\cdot 10^{-15}$. This component of cold, low chloride water contains water originating from the nuclear bomb period (Balderer et al. 2004). It is therefore considered as most representative of a groundwater of recent recharge containing no component of the deep thermal water.
- The water of three deep wells may be considered as the possible end-member of the hot thermal water ascending along deep fault zones. The water of Széchenyi-spa (OBP24), Paskál-spa (OBP 12), and Palatinus-spa Margitsziget II. (OBP1) is tritium-free (<0.5 TU), which have the highest temperature (68–78 °C) and the lowest ^{36}Cl/Cl—ratio of $9.2 \cdot 10^{-15}$–$19.4 \cdot 10^{-15}$. These wells lying on the Pest Plain provide groundwater from limestone and dolomite aquifers covered by thick (300–1,700 m) aquitard sediments.

Table 2 Results of chemical and isotope analyses of the investigated thermal waters

Sample location	Nr.	Type	$^{36}Cl/Cl$ * (10^{-15})	Tritium (TU)	T (°C)	F^- (ppb)	Cl^- (ppb)	Br^- (ppb)	NO_3^- (ppb)	SO_4^{2-} (ppb)
Budapest area										
Széchenyi-spa II	OBP 24	W	9	<0.5	76.8	3.31	189	0.55	n.d.	195
Paskál-spa	OBP 12	W	19	<0.5	69.7	2.69	96	0.25	n.d.	148
Palatinus spa, Margitsziget II	OBP 01	W	17	<0.5	68.6	2.81	166	0.51	n.d.	202
Lukács-spa IV	OBP 26	Sp/W	31	1.5	52.5	2.29	129	0.29	n.d.	161
Csepel-spa II	OBP 20	W	26	<0.5	45.2	1.71	186	0.43	n.d.	321
Gellért-spa GT-I	OBP 15	Sp/W	24	1.5	44.5	2.23	151	0.33	1.2	366
Rudas-spa, Juventus	OBP 06	Sp/W	30	1.3	42.7	2.23	148	0.38	n.d.	283
Dagály-spa, Béke-well	OBP 11	W	23	0.6	37.5	1.34	75	0.24	n.d.	110
Palatinus spa, Margitsziget IV	OBP 03	W	25	0.5	36.9	1.35	76	0.23	n.d.	120
Csillaghegy-spa, József	OBP 10	W	199	1.0	27.8	0.40	18	n.a.	1.7	71
Lukács-spa, Római-spring	OBP 29	Sp	358	4.3	20.8	0.48	34	n.a.	6.3	75
Surrounding area of Budapest										
Csobánka, Szentkút-spring	OBP 21	Sp	1,007	4.3	10.0	0.09	14	n.a.	n.a.	247

The analytical error of the ^{36}Cl measurement made by the Accelerator Mass Spectrometry (AMS) Technique is in the order of 10–20 %. The anion concentrations were determined by ion-chromatography

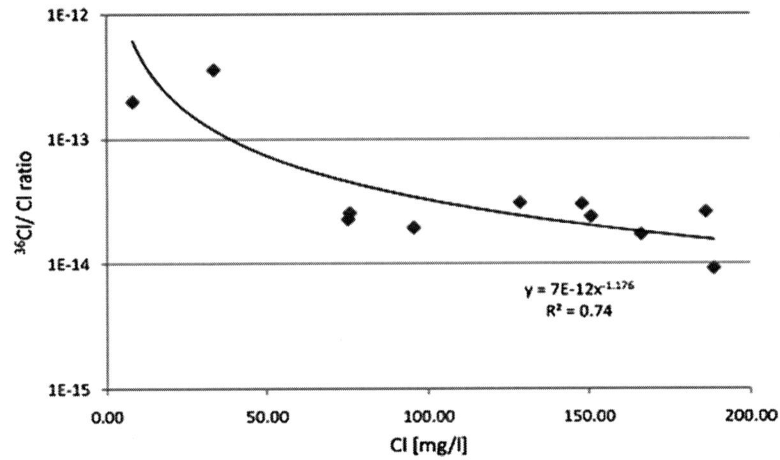

Fig. 4 ^{36}Cl/Cl ratio and Chloride concentration of investigated thermal waters of Budapest

- The other investigated wells and springs are at the vicinity of the fault zone; therefore, these are mixtures of the above-mentioned two components.

From the presented results in Table 2 it can also be deduced that there is no clear systematic trend for the analyzed chemical parameters such as fluoride, chloride, bromide, nitrate and sulfate. This may be due to the fact that the chemical compounds are more related to the local mineralogical composition of the rocks within the groundwater flow path and not only to origin and related mixing processes within the hydrodynamic flow field. For this purpose the two considered radioisotopes present a great advantage, as only related to atmospheric input and groundwater residence time.

4 Evaluation of the Mixing Components of the Budapest Thermal Water

According Eq. (1) the concentration N of atoms of ^{36}Cl in 1 liter of water is determined as: $N = 1.7 \cdot 10^{19} \quad R \cdot C$;
where, R is the measured ^{36}Cl/Cl ratio and C is the concentration of chloride of the investigated water.

The same relationship is represented by the diagram of the ^{36}Cl/Cl ratio against the chloride concentration of the investigated waters (Fig. 4). For this relation results in a regression coefficient r^2 of 0.74, the resulting relative mixing proportions by the application of this relationship are presented in Table 3. The next possibility to test this mixing relation of shallow karst groundwater with deep hot thermal water is the combination with the tritium data.

Table 3 Resulting relative mixing proportions as resulting from the ^{36}Cl concentration

Sample Nr.	N (*10^7at/l)	Mixing proportion of shallow water component
OBP 1	4.86	0.09
OBP 3	3.27	0.01
OBP 6	7.62	0.21
OBP 11	2.89	0.00
OBP 12	3.15	0.01
OBP 15	6.13	0.15
OBP 20	8.26	0.24
OBP 21	24.70	1.00
OBP 26	6.77	0.18
OBP 29	20.40	0.80
OBP 10	2.87	0.00
OBP 24	2.95	0.00

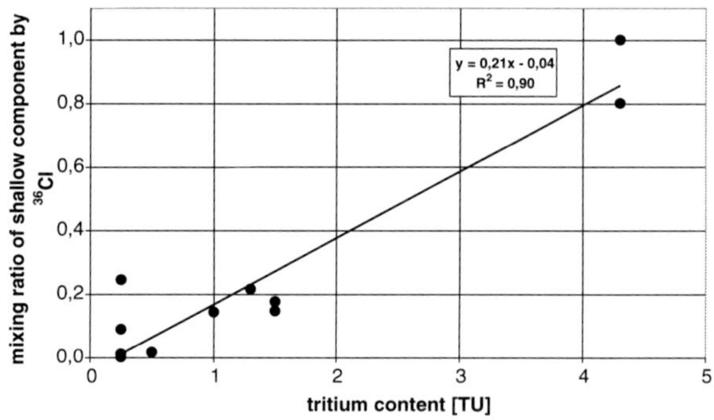

Fig. 5 Mixing ratios of shallow water component as determined from the ^{36}Cl concentration versus the tritium concentration of the investigated Budapest thermal waters

The resulting relation of these calculated mixing ratios with the corresponding tritium concentration (Fig. 5) is with $r^2 = 0.9$ quite good, although some uncertainties may be introduced by the fact that the tritium concentrations in the atmospheric cycle are quite low as the elapsed time as the atmospheric nuclear test activities is now more than 47 years, which equals about four times the half live of the tritium isotope.

The correlation of the $^{36}Cl/Cl$ ratio with the measured water temperature is represented in Fig. 6. Wells and springs in Table 2 are divided into two groups by their tritium content. In the first group where tritium could be detected (>0.5 TU, gray squares on Fig. 6) the $^{36}Cl/Cl$ ratio is decreasing via water temperature. This result can be interpreted as the cooler waters contain more post-bomb ^{36}Cl than the warmer ones because the ratio of fresh (coldest) component decreases via temperature.

In the other group containing tritium free waters (<0.5 TU) the presence of post-bomb ^{36}Cl can be excluded. Significantly lower $^{36}Cl/Cl$ ratio in the warmest waters ($9 \cdot 10^{-15}$–$19 \cdot 10^{-15}$) refers to very long transit time of groundwater or mixing of such an old component to the flow regime. Previous studies (Alföldi et al. 1978; Deák 1979) have demonstrated that the transit time of the Budapest

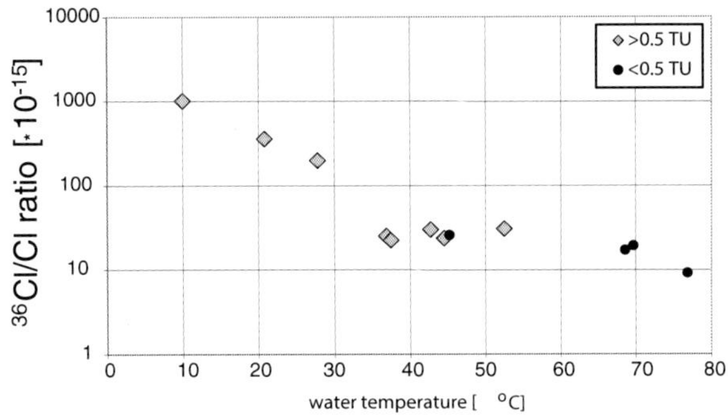

Fig. 6 Correlation of the $^{36}Cl/Cl$ ratio with the measured water temperature of the Budapest thermal waters

thermal karst-water regime is 20–30 thousand years. Accepting these results based on ^{14}C, $\delta^{18}O$ and $\delta^{2}H$ data, mixing of a very old component is more reliable than the extremely long transit time.

5 Conclusion

Based on the results of ^{36}Cl and tritium analyses as presented, the already established conceptual hydrogeological model (Fig. 2) can be confirmed:

- Cold groundwater of atmospheric origin infiltrates within the karstified recharge areas of Buda hills and is flowing within the limestone aquifer—according to the present hydraulic conditions—towards the Danube River which is parallel to an active tectonic fault zone. It supplies the cold component of springs rising along the hill slope. This cold water contains tritium and ^{36}Cl of anthropogenic origin.
- Thermal component originates from the ancient precipitation fallen in the Buda hills and Pilis mountains lying 100–400 m higher elevations as the thermal wells and springs. This high potential energy forces the karst water to follow the karstified rocks into high depths. Reaching the highest depths of the flow path (boundary of carbonate aquifers, see Fig. 2) the heated water will flow towards the springs.
- The deep hot water is characterized by a higher residence time (without any trace of anthropogenic tritium or ^{36}Cl) with $^{36}Cl/Cl$ -ratios at levels quite similar to the ratios of achieved equilibrium of underground production as of vulcanite rocks with $R = 10\ (\pm 1.7) \cdot 10^{-15}$ up to Rhyodacite rocks with $R = 19\ (\pm 1.1) \cdot 10^{-15}$. That would mean that the residence time of the deep water component could even reach time scales of up to 1.5 million years (for achieved equilibrium of underground production).

Such a rock environment is in quite good agreement with the observed geological features of the area whereas even volcanic intrusions of Senonian to Late Eocene age are present. The coeval volcanism produced volcano-clastics, shallow intrusive bodies and dikes. The center of this volcanism was postulated as the Wein palaeovolcano located in the SE foreland of the Buda Hills as reported in Korpás et al. (1999). This hypothetical model is in contradiction with the former interpretation of isotope data (Deák 1979) presenting much younger groundwater ages to be less than 30,000 years. A more reliable hypothetic model can be developed presuming that a certain amount of very old, respectively stagnant (several millions of years old) groundwater is mixed to the relative young (20–30 thousand years) karst water. Mixing ratio depends on the Chlorine ion content of the hypothetic old component.

Acknowledgments This research was partly financed by the Hungarian Academy of Sciences in the framework of the OTKA 60921 project (2007–2010).

References

Alföldi L, Deák J, Liebe P, Lorberer Á (1978) Connection of cold and warm karstic water resources in the Hungarian Central Mountains with special reference to draining efforts of mining (in Hungarian). In: Proceedings of research activity of VITUKI, Report series (1976–1977) 3.15, pp 300–317

Balderer W, Synal HA (1996) Application of the chlorine-36 method for the characterization of the groundwater circulation in tectonically active areas: Examples from North Western Anatolia, Turkey. Terra Nova 8:324–333

Balderer W, Synal HA (1997) Use of chlorine-36 as tracer for the evolution of waters in geothermal and tectonically active areas in Western Turkey. Elsevier Sci 123:387–393

Balderer W, Synal HA, Deak J, Stichler W (1998) Application of the chlorine-36 method for the investigation of the infiltration conditions and evolution of the deep groundwaters of multilayer aquifer systems of the Great Hungarian Plain. In: IAEA-SM-349/2P, vol 2. International Atomic Energy Agency, Vienna, pp 431–437

Balderer W, Synal HA, Deak J (2004) Application of the chlorine-36 method for the delineation of groundwater recharge of large river systems: example of the Danube River in Western Hungary (Szigetköz Area). Env Geol 46(7):755–762

Bentley HW, Phillips FM, Davis SN (1986) Chlorine-36 in the terrestrial environment. In: Fritz P, Fontes JC (eds) Handbook of environmental isotope geochemistry, vol 2, The terrestrial environment, B, Elsevier, Amsterdam, pp 427–480

Deák J (1979) Environmental isotopes and water chemical studies for groundwater research in Hungary. In: IAEA Vienna, isotope hydrology 1978, vol 1, pp 221–249

Korpás L, Lantos M, Nagymarosy A (1999) Timing and genesis of early marine caymanites in the hydrothermal palaeokarst system of Buda Hills, Hungary. Sed Geol 123:9–29

Kovács J, Müller P (1980) Origin and traces of hydrothermal activities in the Buda Range (in Hungarian). In: Karszt és Barlang, 1980/II, pp 93–98

Synal HA, Beer J, Bonani G, Suter M, Wölfli W (1990) Atmospheric transport of bomb-produced ^{36}Cl. Nuclear instruments and methods in physics research, B(NIMB), vol 52. Elsevier, North Holland, pp 483–488

Resources of Curative Mud of the Crimea Peninsula

Elena Kayukova

Abstract This chapter deals with the curative mud resources of salty lakes and mud volcanoes in Crimea Peninsula. There are unique resources of salt lakes brine, salts, curative muds, as well as mineral and thermal water. The medicinal potential of salt lakes of Crimea is very high in which there are 35 salt lakes and 33 mud volcanoes within Kerch Peninsula. Most of the lakes of the Crimea are of marine origin with a permanent infiltration of water from the sea. They are sulphatic type. Total capacity of mud deposits of the Crimea is estimated at about 32,279 mln m^3.

Keywords Crimea Peninsula · Salt lakes · Mud volcanoes · Curative muds

1 Introduction

The Autonomous Republic of Crimea (from 1954 to 1991 it was the Crimean Region) is part of the Ukraine. The Crimean peninsula is in the southern part of Ukraine, washed by the Black and Azov seas and connected to the continent with a narrow Perekop isthmus. The administrative border is located along the Perekop Swell and the Sivash, a shallow bay of the Sea of Azov.

Crimean climate is characterized by its location between the temperate and subtropical geographical zones, contributing to the mild climate of the peninsula and a large number of hours of sunlight (2,180–2,470 h per year). The region is famous for its healing factors: healthful climate, warm sea, mineral waters, mud of salt lakes and volcanoes (that are underestimated). Sometimes the Crimea is called "the small continent," because of its different landscapes and very rich nature. This is also a popular seaside resort; however, the recreation potential of other

E. Kayukova (✉)
Faculty of Hydrogeology, Department of Geology, University of St. Petersburg, Universitetskaya nab. 7/9, St. Petersburg, Russia 199034,
e-mail: epkayu@gmail.com

W. Balderer et al. (eds.), *Thermal and Mineral Waters*, Environmental Earth Sciences,
DOI: 10.1007/978-3-642-28824-1_6,

water resources still has not been studied properly and has not had its complex ecological economic assessment.

During the centuries people have been using natural resources for medical purposes, and curative qualities of many natural factors have been known since ancient times. The Crimean peninsula began to be considered as a perspective resort region after the Empress Catherine II visited in 1787. Later the south coast of Crimea was developed into the most beloved recreation place for Russian aristocracy and leaders (Stalin, Khruschev, Brezhnev, Gorbachov, Yeltsin).

2 Geological and Hydrogeological Settings of the Study Area

The Crimea Peninsula is located at the southwest of the East European Plain. The largest part of its territory has the plain shape. In the west, the plain turns into the lime terraces of the Tarkhankut, running their steep edge into the sea line. The hilly ridges of the Kerch Peninsula dominate the eastern part. The chain of the Crimean Mountains occupies the southern part of the peninsula and tectonically is part of Crimea-Caucasus mountain system.

The Crimea Mountains obtained their shapes in the Alpine fold era. They stretch from west to east for 180 km. There are three ridges: the Main one (the First one, located in the southern margin of the Peninsula) has a height of 1,200–1,500 m; the Interior one (the Second one) has a height of 400–600 m; and the Exterior one (the Third one) has a height of 250–350 m. The Main ridge is the highest and its forestless mountain-massifs are called yayla. Roman-Kosh Mountain (1,545 m) is the highest peak of Crimea located at Babugan-yayla. Coastal hills of the Main ridge end on the southern coast of Crimea.

Geologically, the Crimea Peninsula is situated on the border of the south of the East-European Platform, Skythian plate and Crimea-Caucasian orogenic system. It is subdivided into three parts: southern mountainous part, northern plain part and Kerch peninsula.

Mesozoic and Cenozoic rocks of marine, lagoon and continental origin mainly constitute the geological structure of this region. Within the Main ridge and southern coast, Triassic-Jurassic flysch series and Upper Jurassic—Lower Cretaceous carbonates and terrestrial-carbonates are mostly developed. Within the Second Ridge, Upper-Cretaceous marine clayey carbonates are overlain by Paleogene carbonates. Only Neogene and Quaternary sediments occur in the Pre-Mountainous northern parts and within the Plain Crimea (Muratov 1960; Fig. 1).

The hydrogeological settings of the Crimea Peninsula are different in the northern platform (platform artesian basins) and southern folded part (fissured waters and karstic basins). Kerch Peninsula also has a complex hydrogeology: its southwestern part is an aquiclude: Paleogene Maykop clays 3,000 m thick; its northeastern part is a system of small artesian basins.

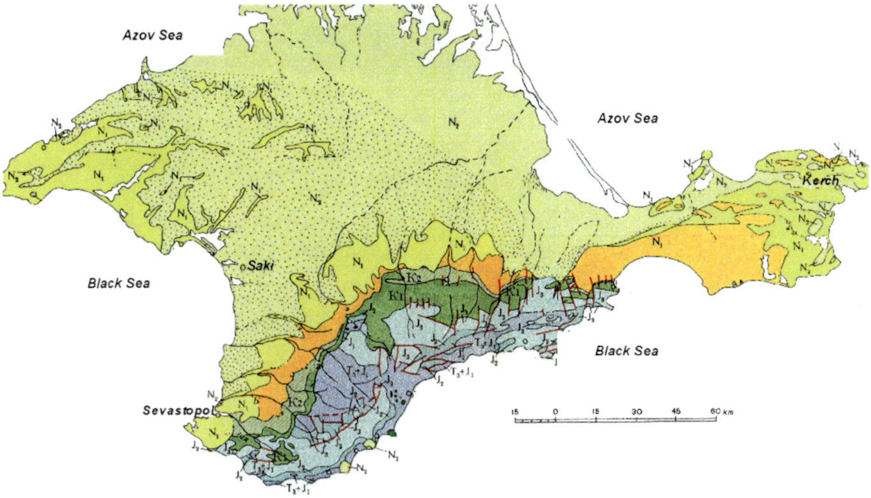

Fig. 1 Geological sketch of the Crimea peninsula (after Muratov 1960). N_2 Pliocene marine sediments; N_2 Pliocene continental deposits; N_1 Miocene deposits; Pg Paleogene deposits; K_2 Upper Cretaceous deposits; K_1 Lower Cretaceous sediments; J_3 Upper Jurassic deposits; J_2 Middle Jurassic deposits; J_1 Lower Jurassic deposits; T_3-J_1 Tavricheskaya Series (Upper Triassic–Lower Jurassic); —— Faults and thrusts

High plateaus of Crimean Mountains (1,200–1,500 m above sea level) are the main recharge area of the whole Peninsula. Groundwater is formed within the Main Ridge, then it comes through surface runoff and underground flow to the north, recharging the artesian basins of plain parts of the region. The close hydraulic connections between the hydrogeological structures of the Crimean Mountains and the northern parts of the peninsula are responsible for the good quality of water in artesian basins, which became an important source of drinking water supply.

The Main Ridge of the Crimea Mountains hydrogeologically belongs to Crimea Mountains' hydrogeological massif, but the Exterior and Interior Ridges are part of the recharge area of Alma and Indolsky artesian basins. Rainfall waters (up to 1,000 mm per year) are accumulated in limestones forming the karstic aquifers.

Apart from the rare landscapes and special natural beauties, nature also endowed Crimea with springs and lakes with important therapeutic properties.

3 Mineral Resources of the Salt Lake and Mud Volcanoes

Mineral resources of the Crimean Peninsula include mineral waters, thermal springs, curative muds in salt lakes and volcanoes and many touristic landmarks. More than 300 salt lakes, lagoons and gulfs are situated along the seacoast. They

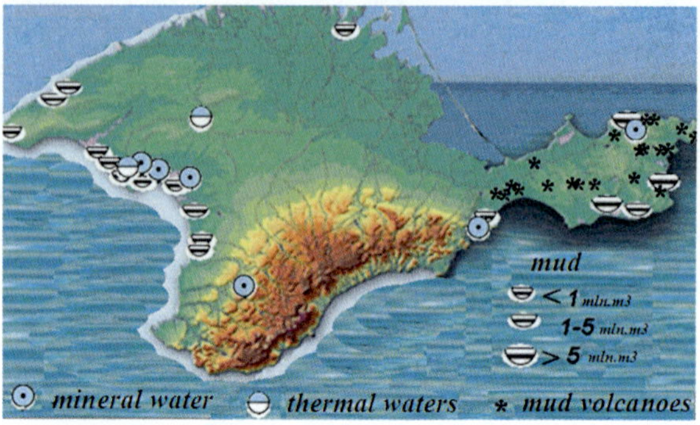

Fig. 2 Position of the main resources of mineral waters and therapeutic muds

contain brines and curative muds; many salt lakes are used for extraction of table or bathing salts.

There are 33 mud volcanoes discovered within Kerch Peninsula (the host rocks are Paleogene Maykop clays). A map of mud reserves and mud volcanoes is presented on Fig. 2.

All lakes in Crimea are divided into 7 subgroups: I—Perekop subgroup, II—Tarkhankut subgroup, III—Evpatoriya subgroup, IV—Yaila lakes, V—Khersones subgroup, VI—Kerch subgroup, VII—Chongar-Arabat subgroup (Fig. 3).

Salt lakes in the Crimea peninsular have the total surface area of about 27,000 km^2. Those with TDS of the water of more than 35 g/l occupy more than 15 % of this area. Only a few of them have continental origin and never had a connection with the sea. The most widespread are lakes of sulphate type water (Ponizovsky 1965). These types of lakes are divided into sulfate-sodium and sulfate-magnesium (or chloride magnesium). The lakes of sulfate-magnesium water of marine origin are dominant types of lakes. Lakes of continental origin are mostly sulfate sodium.

A great majority of the salt lakes are situated along the coast in the steppe parts. Most of them are shallow and sometimes dry up during the summer.

There are 26 lakes where deposits of therapeutic mud and brine as well as over 100 mineral springs of different chemical compositions have been distinguished. According to the decision of Ukrainian authorities, 15 places with mud deposits and 13 large resources of mineral water were granted by therapeutic category (Khmara et al. 2001). That is one of the priorities in the development of the Crimean resort area. Data characteristic features concerning deposits of therapeutic mud are presented in Table 1. There are only three large salt lakes (with an area of more than 10 km^2) in this table, such as Sasyk-Sivash, Uzunlarskoe, Tobechikskoe due to the fact that not all lakes and their mud deposits are of medicinal importance.

Geological reserves of curative mud in Crimea are about 30 mln m^3 (Khmara et al. 2001). At present, the world market price of curative mud is about 5 USD per kg. The most known natural salt lakes are: Sakskoye (near Saki), Moynaki (near Evpatoriya), Sasyk-Syvash (east Evpatoria). Other important lakes

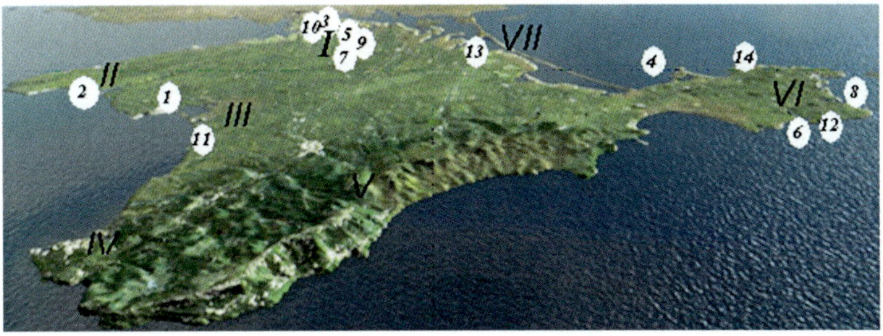

Fig. 3 Position of the main salt lakes of the Crimean peninsula, *1* Sasyk, *2* Donuzlav, *3* Aygulskoe, *4* Aktashskoe, *5* Krasnoe, *6* Uzunlarskoe, *7* Kirleutskoe, *8* Tobechikskoe, *9* Kiyatskoe, *10* Staroe, *11* Sakskoe, *12* Koyashskoe, *13* Genicheskoe, *14* Chokrakskoe

Table 1 Salt lakes and mud volcanoes of medicinal importance (after Khmara et al. 2001)

No.	Deposit		Reserves (m^3)	Recommendation for use
1	Kizil-Yar	mud	10,000,000	Central and peripheral nervous system (neuralgia, neuritis); cardiovascular system; musculoskeletal system, the effects of trauma; diseases of the joints (arthritis, arthrosis, hondrozy); digestive; gynecological diseases; skin diseases
2	*Uzunlarskoe*	mud	6,930,000	
3	*Tobechikskoe*	mud	5,500,000	
4	*Bulganakskoe*	mud	5,000,000	
5	*Chokrakskoe*	mud	4,660,000	
6	Sakskoe	mud, brine	4,500,000	
7	*Koyashskoe*	mud	3,369,000	
8	Dzharylgach	mud	3,100,000	
9	Sasyk-Sivash	mud	976,000	
10	Bagayly	mud	120,000	
11	*Djau-Tepe*	mud	32,000	

Deposits of Kerch Peninsula are emphasized in italics

are: Chokrak, Uzunlarskoe, Koyashskoe, Tobechikskoe (Kerch peninsula). There are a lot of legends about the healing power of these lakes.

The curative muds are divided into two main types depending on the physical and chemical properties: silt sulfide type and Sopochnaya type (hill mud; Ponizovsky 1965; Fig. 4). They are distinguished by smell, texture, temperature, content of organic substances (e.g. the silt sulfide muds contain up to 28 % organic matter whereas the sopochnaya muds contain practically no organic matter).

Sulfide silt muds are rich in iron sulfides and water-soluble compounds. Uzunlarskoe Lake is the largest reservoir of sulfide silt mud in the Kerch region. The potential resources of therapeutic muds of lakes Kizil-Yar, Uzunlar, Chokrak, Koyash, Dzharilgach, Sakskoe and others are very large; however, not all of them are utilized by the spa industry, presumably due to very bad road infrastructure. The first mud-cure resort in Russia was Saki, which was founded in 1827. The

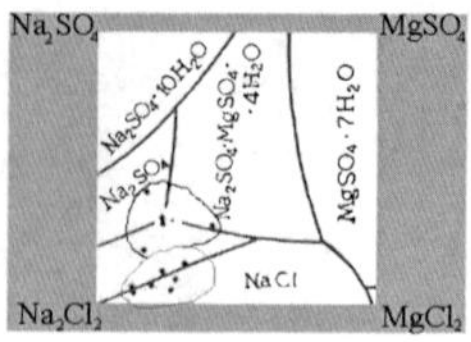

Fig. 4 Diagram showing the chemical composition of salt lakes in the Crimea peninsula (after Ponizovsky 1965)

waters of Saki Lake are recognized as the saltiest in the world with the high concentration of medicinal mud. Now the resources of mud have decreased from 21 to 5.5 mln m^3.

3.1 Silt Sulphide Mud (Lakes of Marine Origin)

Chokrakskoye Lake (marine origin) is a unique reservoir of mud and brines. The mud reserves are estimated at 4,660,000 m^3. The lake is recharged from time to time by seawater, usually during stormy weather, when the water from the sea can reach the lake through the narrow bar that divides both reservoirs (Fomichev 1948; Figs. 5 and 6). On the bottom of the lake there are mud volcanoes, mineral springs and gas emanations. Chokrak Lake mud is of sulfate–chloride-magnesium-sodium type with high sulfides content. The mud contains: hydrogen sulfide, iron sulfide, nitrogen, hydrogen, hydroxides of aluminum and iron salts of fatty acids, traces of precious metals (e.g. silver, gold, platinum), bitumen-like compounds, lignin, cellulose, organic retinoid substances, most of the known amino acids, fulvic acids, organic acids, aromatic derivatives, various biologically active substances, vitamins, biogenic stimulants, substances such as antibiotics, and the entire set of sea salt.

New substances and chemical compounds are formed also as a result of algae and other micro-organism activity. Chokrak muds are counted among those of high therapeutic activity and are used for healing of many diseases.

Chokrak Lake is located in the northern part of Kerch Peninsula. The water in this lake has TDS of about 80–300 g/l, depending on the season: Cl^-—72,500 mg/l; SO_4^{2-}—38,000 mg/l; HCO_3^-—1,100 mg/l; CO_3^{2-}—300 mg/l; Ca^{2+}—300 mg/l; Mg^{2+}—24,700 mg/l; $(Na+K)^+$—60,300 mg/l; pH—6.8. The brines are chloride magnesium-sodium. They contain up to 750 mg/l of bromine and up to 300 mg/l of boron (as H_3BO_3).

A group of hydrogen sulfide springs is located on the eastern edge of the lake. The water contains iodine, bromine, boron, iron, titanium, aluminum, barium, manganese, copper, strontium, and lithium. The waters are bicarbonate-chloride-sodium and chloride-sodium.

Koyashskoye Lake (marine origin) is located in the southern part of the Kerch Peninsula within the Opuksky Nature Reserve and Koyashskoe Lake (Fig. 7). Brines of Koyash Lake contain sodium chloride, magnesium and potassium chlorides, sodium iodide, magnesium bromide and magnesium sulfate, calcium

Fig. 5 General view on Chokrakskoe lake of marine origin

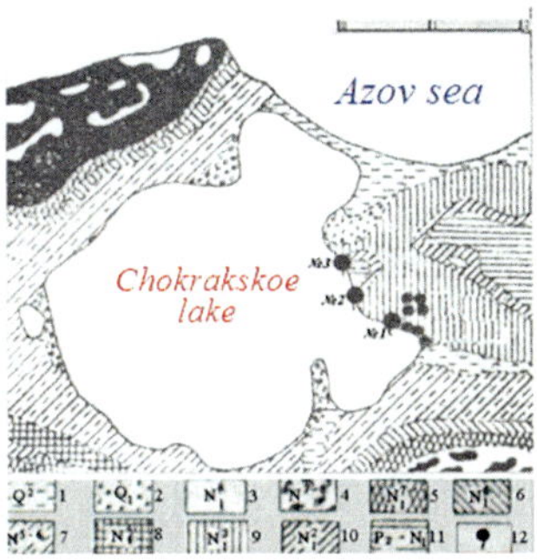

Fig. 6 Geological sketch of lake Chokrakskoe lake area (after Fomichev 1948). Explanations, *1* Coastal sands, loess loam, deposits of the lake Chokrak, *2* Caspian and mediterranean terrace, *3* Meotian, *4* Upper Sarmatian and Meotian reefs, *5* Upper Sarmatian, *6* Medium Sarmatian, *7* Middle and Upper Sarmatian clays, *8* Konksky horizon, *9* Karagan horizon, *10* Chokrak horizon, *11* Maikop series, *12* Springs

Fig. 7 General view on Koyashskoe lake

sulfate and organic matter. The brines of Koyash Lake have Cl^-—107,500 mg/l; SO_4^{2-}—19,600 mg/l; HCO_3^-—610 mg/l; Ca^{2+}—200 mg/l; Mg^{2+}—4,480 mg/l; Na^+—69,000 mg/l; K^+—3,900 mg/l. In the summer season the lake salinity is about 200–250 g/l and more. The reserves of high-quality curative mud are more than 3 mln m^3. Valuable compounds in lake waters are mainly potassium, iodine, bromine, boron, arsenic, antimony, mercury, and gold.

One of the interesting features of Koyashskoe Lake is the pink color of the salt crystals formed after substantial evaporation of the lake water. Such color of the

Fig. 8 Crustacean *Artemia salina* (Koyashskoe lake)

salt crystals indicated the presence of microalgae *Dunaliella Sallina* (Fig. 7). This species, the representative of the plankton community, is the main supplier of provitamine A and products of its transformation in muds, which increases the mud's therapeutic properties. Microalgae *Dunaliella* serves as food for crustacean *Artemia Sallina,* which can live in a highly saline environment up to 250 g/l (Fig. 8). The resources of Koyash Lake are not utilized for curative purposes because of its remoteness and bad roads.

3.2 Silt Sulphide Muds (Lakes of Continental Origin)

The lakes of continental origin (i.e. salt pans) occur in falls of a relief and are not connected with the sea. The local name of such lakes is 'koli' (Fig. 9). Such lakes are formed mainly from waters of atmospheric precipitation. Almost all of them are located on the southwestern plains of the Kerch Peninsula. They are not large and are rather shallow. During the summer season they dry out partially or completely, and a larger area is covered by the mineral crust of white, white and pink or white–gray color, under which there is a very thin layer of brine. This layer is underlain by silts, mostly black or gray, with a significant water content. The chemical composition of the lake water is different from the seawater and is of sulfate type.

3.3 Sopochnaya Mud (Hill Mud)

The mud volcanism on the Kerch Peninsula is manifestation of modern tectonic. There are 33 mud volcanoes on the Kerch Peninsula, and 7 in the Black and Azov Sea (Fig. 10). At present, activity of mud volcanoes is not large; they produce cold

Fig. 9 General view on the Marfovka salt lake of continental origin

Fig. 10 Sketch presenting the position of mud volcanoes of the Kerch peninsula (after Shnyukov et al. 1992)

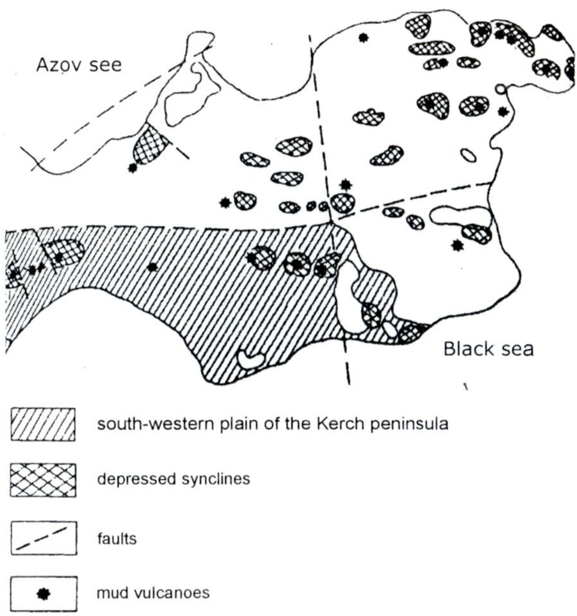

mud, water, and emit gases (mostly methane and carbon dioxide). Sometimes powerful eruptions occur, and the volume of produced mud reaches significant values.

Most interesting are volcanoes located at the Bulganakskoe field (Fig. 11). The reserves of the Bulganak muds are estimated at 5,000 m^3. Groundwater of mud volcanoes is usually saline; their TDS reaches 32.6 g/l with ion concentration as follows: $Cl^- = 5{,}453$ mg/l, $SO_4^{2-} = 1{,}920$ mg/l, $HCO_3^- = 15{,}128$ mg/l, $Ca^{2+} = 24$ mg/l, $Mg^+ = 29$ mg/l, $Na^- = 10{,}074$ mg/l, and pH = 9.

Volcanic mud is a slurry of fine solids suspended in water. The main components of mud are: silica—54.9 %, alumina—15.7 %, iron—6.9 %, water—4.1 %. Other components that can be found in mud are: chlorine, iodine, bromine, fluorine, boron (up to 0.21), lithium, strontium, phosphorus, ammonium, manganese,

Fig. 11 Examples of the mud volcanoes of Bulganakskoe field

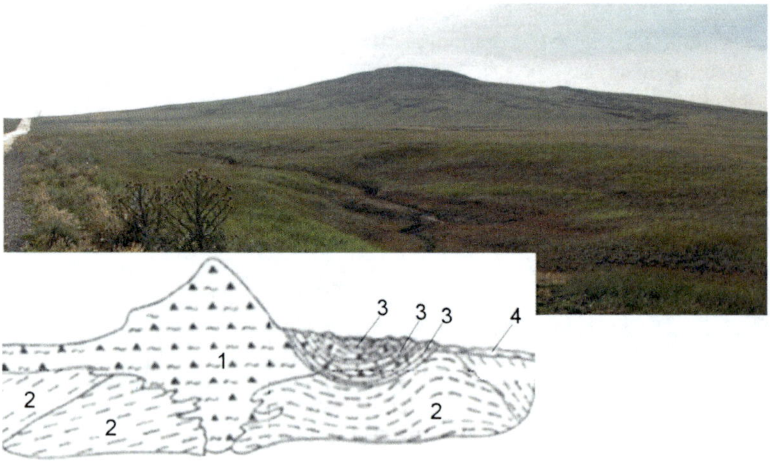

Fig. 12 View on the Djau-Tepe mud volcano and generalized geological cross section (after Shnyukov et al. 1992). Explanations, *1* Sopochnaya formation (clays, slurry, silts, sands), *2* Maykop clays, *3* Limestones, *4* Quaternary sediments

titanium, nickel, cobalt, vanadium, zirconium, copper, zinc, lead, tin, chromium, barium, beryllium, scandium, gallium, arsenic, mercury, silver, organic matter (humic acids, bitumens, volatile fat acids, amine compounds and carbohydrates), and microflora (dies in contact with air). Weak radioactivity and alkaline media are typical features of these muds. In the gas phase methane, hydrogen sulfide and helium are present. The mud is oily and elastic, almost sterile, and has good balneological properties.

Mud volcanoes are located usually in areas of tectonic activity and where deposits of oil and gas occur at depth. The mud volcanoes on Kerch peninsula are located on the vault of Vulkanovskoy anticline that extends in the latitudinal direction. According to available geological data, the roots of volcanic structures can reach Lower Cretaceous and Jurassic strata. Mud volcanoes are associated with bulged synclines. There is a thick layer of Maikop clay with a considerable quantity of organic matter. In conditions of elevated pressure and temperature the organic matter is decomposed with the release of hydrocarbons (mainly methane).

When the pressure of released gas rises, it goes to the surface through privileged ways pushing ahead fine solids suspended in water. When the pressure drops, the activity of the volcano stops.

The largest mud volcano on the Kerch peninsula is Djau-Tepe (Fig. 12). The height of the hill is 116 m above the sea level. Ejections of Djau-Tepe gases consist of heavy hydrocarbons (83 %) and carbon dioxide. However, most of the mud volcanoes of the Kerch peninsula eject methane. Sulfuric springs occur on the southern side of the Djau-Tepe. The specific components in the spring water are: B—62.2 mg/l, As—0.01 mg/l, Li—0.4 mg/l, P—1 mg/l, Rb < 0.07 mg/l, Sr—2.1 mg/l, F—0.2 mg/l, Br—5.3 mg/l, I—5 mg/l, Cs < 0.05 mg/l (Lagunova 1972).

The mud deposits of Djau-Tepe occupy an area of about 1.5 km^2. The total volume of the mud ejected by this volcano is calculated at 55 million m^3. The reserves of the mud are estimated at 32,000 m^3.

4 Conclusions

Most of the lakes of the Crimea peninsula, especially the largest ones, are of marine origin with a permanent recharge of water from the sea. The majority of lakes are sulphatic type. All the lakes are characterized by high seasonal variation in volume, surface area, salinity and water temperature.

There are many problems in utilization of the therapeutic resources of the salt lakes and mud volcanoes connected with natural and social reasons (e.g. bad road network, remoteness of lakes, etc.). To date many salt lakes of the Crimea peninsula have disappeared (e.g. from 48 lakes to 35 presently); the chemical composition of water in some lakes is changing continuously due to antropogenic contamination (e.g. Lake Donuzlav).

Crimea peninsula is one of the most unique places in the world with respect to occurrence of salt lake brines, curative mud resources, and other salts and chemical compounds, which can be extracted from brines or muds. The muds contain components formed in long-lasting geological, biological, chemical and physical processes. The medicinal and therapeutic potential of salt lakes of Crimea is very high and even exceeds similar products of the Dead Sea.

Total capacity of curative resources (mud and brines) of the Crimea (34 mud deposits and lakes) is: brines—91.2 mln m^3, therapeutic muds—32.279 mln m^3 (data for 1997).

Before 1999, the curative mud was extracted about 15,000 m^3 from salt lakes (Saki and Chokrakskoe). Practically unlimited resources, therapeutic properties and unique quality of the salt lakes brines and muds allow the Ukraine to create health-care and resort facilities of international significance.

References

Fomichev M (1948) Chokrakskoe hydrogen sulfide sources. In: Savarenskii F (ed) Proceedings of the laboratory of the hydrogeology problems, T. 1, USSR Academy of Sciences, pp 221–232 (in Russian only)

Khmara A, Khlebnikov A, Ivanova V, Dyakovich P, Kapinos N (2001) Mineral resources of Crimea and the adjacent waters of the Black and Azov Seas. Atlas: the application to the collection *Problems of development of Crimea.* Simferopol, Tavria—plus, 81 p (in Russian only)

Lagunova I (1972) The chemical composition of mud volcanoes groundwater of the Kerch-Taman region. Dissertation. LSU, 161 p (in Russian only)

Muratov M (1960) A brief sketch of the geological structure of the Crimean peninsula, 208 p (in Russian only)

Ponizovsky A (1965) Salt resources of Crimea. Simferopol, 263 p (in Russian only)

Shnyukov E, Gnatenko G, Nesterov V, Gnatenko O (1992) Mud volcanoes of the Kerch-Taman region. Kiev: Nauk. Dumka, 200 p (in Russian only)

State of the Environment and Natural Resources (Geological Environment) (1998) Issues of development of Crimea, Issue 10, 180 p (in Russian only)

Chemical and Isotopic Characteristics of Thermal Waters in the Carpathian Region, South Poland: Implication to the Origin and Resources

Adam Porowski

Abstract The Polish part of the Carpathians is the region in which thermal water potential is still under investigation and exploration, mainly due to complex geological structure and variable hydrogeological settings. Excluding the quite well recognized area of the Podhale artesian basin, the huge area of the so-called Outer Carpathians is poorly recognized with respect to thermal water occurrence, resources and possibilities of their utilization. In the current study, to shed some light on the occurrence and origin of potential resources of thermal waters in different parts of the Carpathian Geothermal Province, five exemplary thermal waters have been taken into consideration. Their chemical and isotopic compositions have been shown in the light of the hydrogeological and thermal conditions of their occurrence. The Podhale basin, located within the Inner Carpathian Geothermal Region, is the best recognized with respect to thermal water occurrence. The main horizons of thermal waters occur at a depth of 1,000–3,500 m and are linked to Eocene carbonates and Middle Triassic limestones and dolomites. The meteoric origin of thermal waters, vicinity of the recharge area, low mineralization of waters ($\leq$2.6 g/dm^3) and favorable thermal conditions (outflow temperature of waters in the lower part of the hydrogeothermal systems reaches 86 °C) make this region the most perspective and the most developed with respect to thermal waters utilization. The Outer Carpathians Geothermal Province is composed entirely of flysch-type sedimentary rocks from Late Jurassic (western part) and Early Cretaceous (eastern part) to Oligocene. The characteristic feature of the Outer Carpathians is a complex, folded and thrusted structure. The distinction of individual hydrogeothermal systems is hardly possible. The thermal waters are usually brackish, saline and brine; in the eastern part they are associated with oil and gas deposits. The outflow temperatures usually do not exceed 45–50 °C and well discharges are low. The recharge areas for thermal waters are still not known;

A. Porowski (✉)
Polish Academy of Sciences, Institute of Geological Sciences, Twarda 51/55, 00-181 Warszawa, Poland
e-mail: adamp@twarda.pan.pl

W. Balderer et al. (eds.), *Thermal and Mineral Waters*, Environmental Earth Sciences, DOI: 10.1007/978-3-642-28824-1_7,

their origin is poligenetic and their resources might be limited. Also the thermal conditions in the area of flysch Carpathians seem to still be poorly recognized, reveal variable character but locally might be very promising.

Keywords Carpathian mountains · Thermal water resources · Isotopic composition · Chemical composition · Geothermometers · Hydrogeological settings

1 Introduction

The main reservoirs of thermal waters in Poland occur within the sedimentary basins of Polish Lowlands and crystalline massifs of the Sudetes Mountains. The Polish part of the Carpathian Mountains is the region in which the thermal water potential is still under investigation and exploration, mainly due to complex geological structure and variable hydrogeological settings. Excluding the quite well recognized area of the Podhale artesian basin, located within the Inner Carpathians, the huge area of the so-called Outer Carpathians is poorly recognized with respect to thermal waters occurrence, resources and possibilities of their utilization. This is the main cause that the Carpathian Geothermal Province as a whole unit is usually omitted in many important publications and studies concerning the resources of thermal water in Poland.

It is well known that the Polish part of the Carpathians Mountains abounds with mineralized waters utilized in many spas for balneological treatment. Many aspects of the occurrence of such waters, chemical and isotopic composition, resources and origin have been investigated and discussed widely in literature, for example: Dowgiałło (1976, 1980), Dowgiałło and Sławiński (1979), Dowgiałło and Leśniak (1980), Leśniak (1980), Zuber and Grabczak (1985, 1986, 1987), Chowaniec (1989, 2009), Chowaniec et al. (2001), Porowski (2001a, b, c, 2004a, b, 2006). Thermal waters, however, are best recognized only in the Podhale Basin where, to date about 12 working thermal wells exist (for example: Chowaniec and Poprawa 1985, Sokołowski 1985; Chowaniec 1989, 2009; Kępińska 1995, 2001; and many others cited by these authors). Within the Outer Carpathians thermal waters have been encountered in several drillings—in the majority of exploratory ones for oil and gas, which were liquidated if missed (Chowaniec et al. 2001). Only a few drillings have been made for exploration of thermal waters. Some of them, like those in Ustroń, exist to date as working thermal water wells.

This chapter is focused especially on the comparison of chemical and isotopic characteristics of exemplary groundwaters of temperature at the outflow higher than 20 °C, which are encountered in the Polish part of the Carpathians. The main objective is to show the variation of chemical and isotopic composition of thermal waters induced by variable hydrogeological, and geochemical conditions in different parts of the Polish Carpathians. The chemical and isotopic data presented in the chapter are the results of the field survey performed by the author in the years 2003–2006.

2 Hydrogeological Settings and Thermal Water Occurrence

The Polish part of the Carpathian Mountains, which constitutes the so-called Carpathian Geothermal Province, consists of two geologically distinct parts: (i) the Inner Carpathians—the southern part, composed of crystalline rocks and Mesozoic sediments, folded during the Late Cretaceous; and (ii) the Outer Carpathians (Flysch Carpathians)—the northern part, composed of Cretaceous and Paleogene sedimentary formations, folded during the Neogene (Fig. 1). The Inner Carpathian Geothermal Basin and the Outer Carpathian Geothermal Basin are completely different with respect to geology, hydrogeology, origin and occurrence of mineral and thermal water resources.

2.1 The Inner Carpathian Geothermal Basin

The Inner Carpathians are composed of three geological units: Tatra Mountains in the south, Pieniny Klippen Belt in the north and the Podhale Basin in between (Figs. 1 and 2). The Pieniny Klippen Belt composed of Mesozoic carbonate sediments and volcanic rocks, is a long boundary structure, which separates Inner and Outer Carpathians. The Podhale Basin is the best recognized and is considered as the most perspective geothermal area within the whole Polish Carpathians (Sokołowski 1985; Chowaniec 2009).

The basin is composed of Eocene–Oligocene flysch sediments (mainly sandstones and shales), which cover the Eocene carbonates: mainly nummulitic limestones, conglomerates and mudstones (Fig. 2). The maximum thickness of flysch complex reaches about 3,000 m and is considered to be the cover for thermal water bearing horizons. Both series cover the sub-Tatric nappe structures consisting of Mesozoic sediments and extending up to the Pieniny Klippen Belt. The main horizons of thermal waters occur at a depth of 1,000–3,500 m and are connected with deep, regional groundwater flow system in Eocene carbonates in the upper part and Middle Triassic limestones and dolomites, sometimes Jurassic sandstones, in the lower part (Kępińska 2001). Particular aquifers are separated by layers of very low permeability, and waters can mix only in zones of tectonic engagement. The water bearing horizons dip northward and form typical artesian aquifers. The local groundwater circulation system contains fresh waters and is connected with Quaternary sediments and upper parts of Eocene—Oligocene flysch. Within the Polish part of the Podhale Basin more than 12 thermal wells have been drilled to date. The shallowest well, Zazadnia IG-1, 680 m depth, has a water outflow temperature of about 22 °C; the deepest one, Banska IG-1, 5,261 m depth, −82 °C.

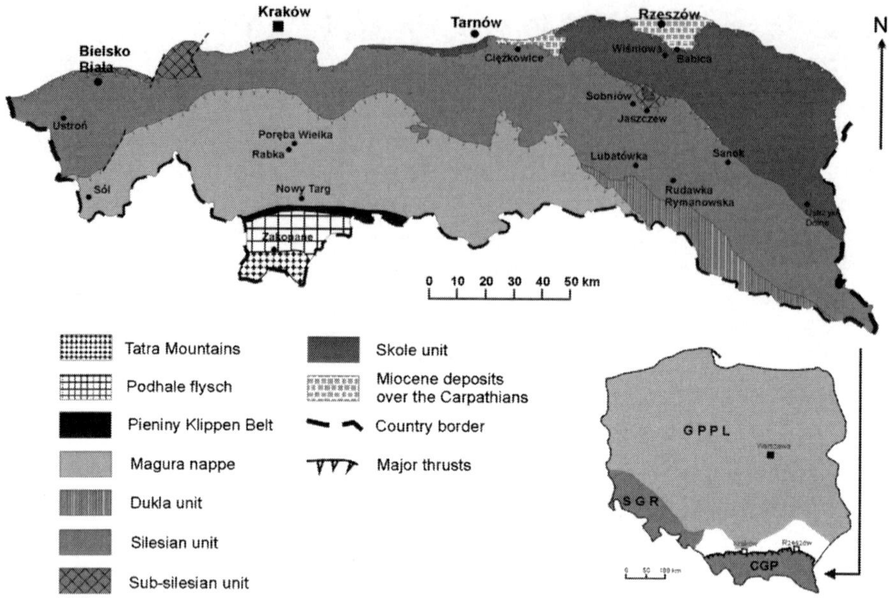

Fig. 1 Generalized structural sketch of the Carpathian Geothermal Province (*CGP*) and its position in Poland. *GPPL* Geothermal province of polish lowlands, *SGR* Sudetic geothermal region

2.2 The Outer Carpathian Geothermal Basin

The Outer Carpathians are composed entirely of flysch-type sedimentary rocks deposited in a deep sea (Książkiewicz 1972, 1975; Kuśmierek 1995). Dominant are detritic rocks as claystones, mudstones, sandstones, conglomerates, and shales. The sedimentary sequence shows a continuity of marine sedimentation from Late Jurassic (western part) and Early Cretaceous (eastern part) to Oligocene. The characteristic feature of the Outer Carpathians is a complex, folded and thrusted structure. They are composed of several nappes (referred to as units in geological literature) that were thrusted over one another approximately from the south (Fig. 1). These nappes are: (i) the Magura Nappe (ii) the Dukla Nappe (iii) the Silesian Nappe (iv) the Subsilesian Nappe, and (v) the Skole Nappe. The nappes differ in tectonic style and lithostratigraphic inventory as a consequence of regional differentiation of sedimentary systems and the distance from the alimentation zones. One of the largest and best developed is the Silesian Nappe.

The easternmost part of the Silesian Nappe constitutes a large tectonic structure called the Central Carpathian Synclinorium built of the thickest sequence of flysch dominated by coarse-grained sediments of the Upper Cretaceous to Oligocene (Fig. 3).

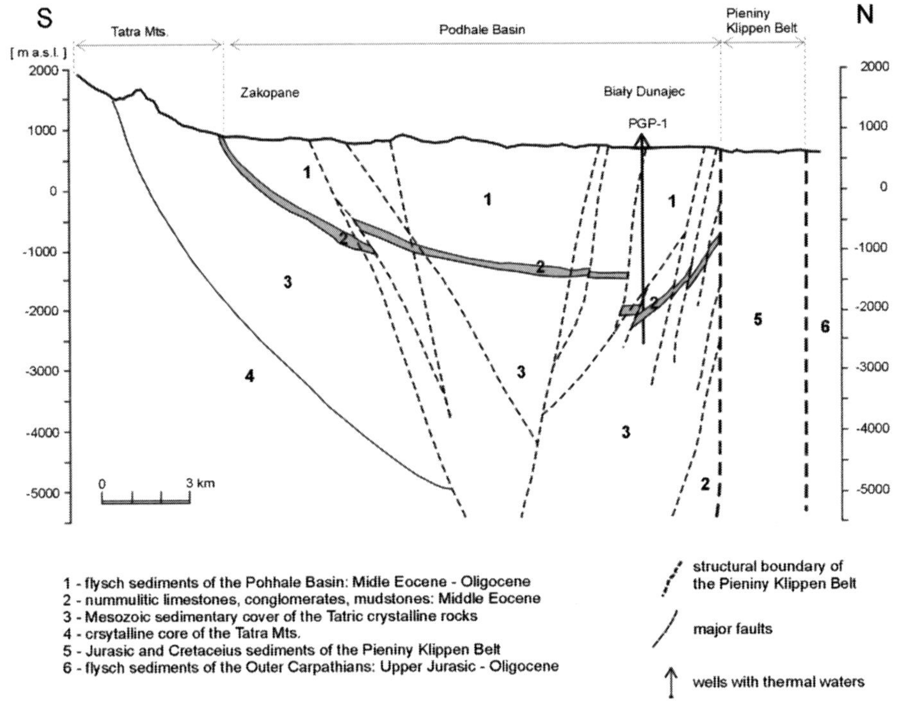

Fig. 2 Schematic geological cross section through the Podhale Basin (after Chowaniec 2003) and the approximate position of studied thermal well

Within the Central Carpathian Synclinorium from seven to eight thrusted anticlines (folds) are distinguished (Ślączka 1977). They are extended from WNW to ESE and usually are tens of kilometers long and only up to a few kilometers wide. The folds are considerably faulted perpendicularly to their axes, which also undulate vertically. Almost all of these folded structures in their ridge parts contain hydrocarbon deposits and associated mineralized waters. The oil-associated or edge oil-field waters sometimes reveal quite high capacities, especially in the wells where resources of oil or gas are running down fast, and have temperatures at the outflow higher than 20 °C, which creates some prospects for their future utilization as thermal resources. The Upper Cretaceous and Tertiary flysch sediments are major constituents of the Central Carpathian Synclinorium and eastern part of the Silesian Nappe whereas, the most important sedimentary formations providing collectors of hydrocarbons and mineralized thermal waters in this area are: (i) the Upper Cretaceous—Palaeocene Istebna beds (sandstones and shales); (ii) the Upper Palaeocene—Lower Eocene Ciężkowice sandstones and variegated shales, and (iii) the Oligocene Krosno beds. The number of sandstone-shale sequences and their thickness in different fold structures vary considerably.

In this part of the Outer Carpathians thermal waters have been encountered in exploratory drillings also in the vicinity of Ciężkowice, Rudawka Rymanowska,

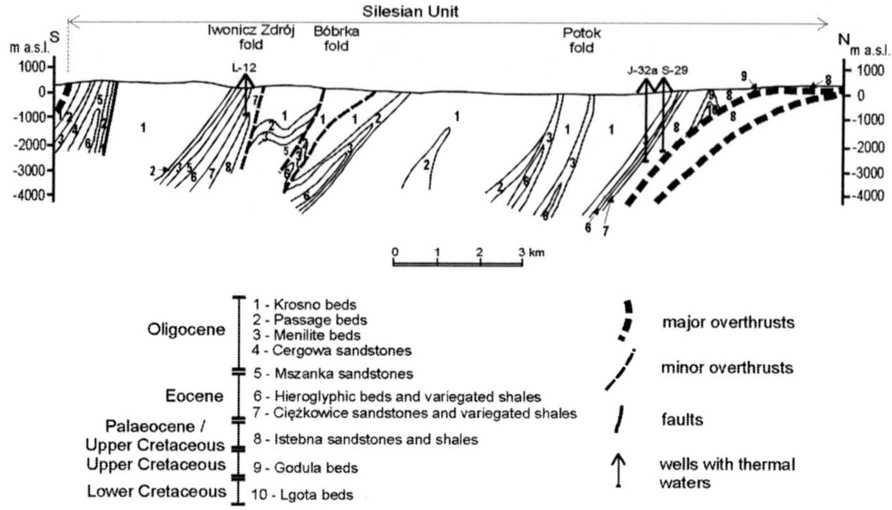

Fig. 3 Schematic geological cross section through the Central Carpathian Synclinorium (based on Neścieruk et al. 1992) and approximate position of the studied wells: *J-32a* well Jaszczew 32a, *S-29* well Sobniów 29, *L-12* well Lubatówka 12

Wetlina and Polańczyk; in Skole Unit—in the vicinity of Wiśniowa and Babica (Fig. 1). In all of these places the thermal waters were saline and occurred in flysch sediments. Maximum outflow temperature of 84 °C was found in water extracted from Wiśniowa-1 exploratory borehole, where water bearing horizons occurred in Lower Cretaceous sandstones at depth intervals 3,696–3,698 and 3,790–3,793; surprisingly the water had quite a low value of TDS in the range of 7 g/dm^3.

The western part of the Outer Carpathians, and Silesian Nappe as well, are formed of flysch sediments of the Upper Jurassic to Oligocene. The Cieszyn shales of the Upper Jurassic—Lower Cretaceous age are the oldest and Oligocene Krosno beds are the youngest (Ryłko and Paul 1998). In this part of the Outer Carpathians the flysch sediments reveal much lower thickness in comparison to the eastern part. The Silesian and Subsilesian Nappes are overthrusted on the formation of Carpathian Foredeep Miocene, which is composed mainly of illitic shales, sandstones and conglomerates (Fig. 4). Miocene formation covers directly the Paleozoic sediments represented by Upper and Lower Carboniferous sandstones, mudstones and clayey shales, and Upper and Middle Devonian carbonates. Paleozoic sequence directly covers the Precambrian crystalline basement of the Carpathians, which is composed mainly of gneisses. Contrary to the eastern part of the Outer Carpathians, the most perspective formations with respect to thermal water occurrence are those underlying flysch sediments, namely the Lower Carboniferous—Devonian carbonates (Sokołowski and Kempkiewicz 2001).

In the vicinity of Ustroń five thermal wells have been drilled from the sixties of the last century. As much as three thermal water bearing horizons have been recognized, namely: (i) the first one at a depth range of 950–1,200 m in Caboniferous

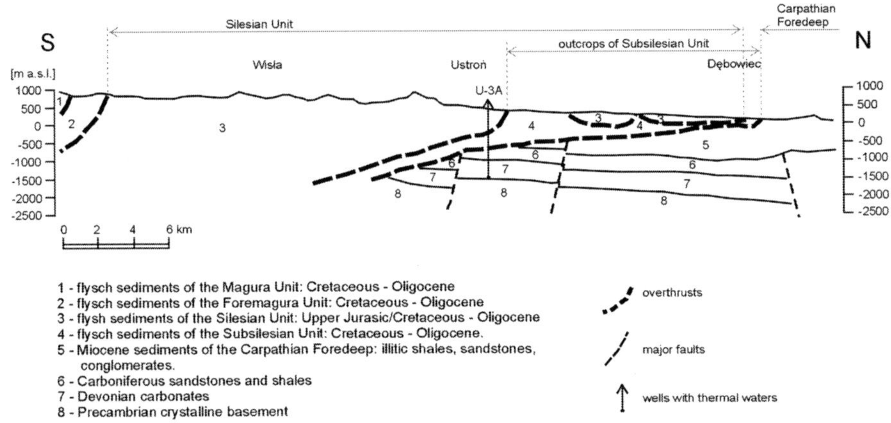

Fig. 4 Schematic geological cross section in the vicinity of Ustroń, west part of the outer Carpathians (after Sokołowski and Kempkiewicz 2001) and approximate position of the studied thermal well

sandstones with average reservoir water temperatures about 46 °C (ii) the second one at depth between 1,200 and 1,700 m in Devonian carbonates with reservoir water temperatures of 45–55 °C, and (iii) the third one—in Precambrian crystalline rocks in the depth interval of 1,700–1,900 m with reservoir water temperatures of 55–65 °C (Sokołowski and Kemkiewicz 2001). However, due to relatively low well discharge (usually <1 m^3/h) the outflow water temperatures are much less, and do not exceed 25 °C.

There are also other places in the western part of the Outer Carpathians where thermal waters have been found. For example, in Rabka spa, located within the Magura Unit (Fig. 1), thermal water with an outflow temperature of about 28 °C are exploited from Cretaceous—Paleocene Ropianieckie beds. In the vicinity of Sóil, saline thermal water has been encountered during exploratory drilling. The outflow temperature of this water reached 38 °C and 18 m^3/h discharge. The water of highest outflow temperature of about 42 °C was found in borehole IG-1 in Poręba Wielka located also in the area of Magura Unit; the thermal water aquifer occurred in Oligocene Krosno beds.

3 Chemical and Isotopic Composition of Typical Thermal Waters

To show large variation among thermal waters encountered in the Carpathian Geothermal Province typical waters from five working wells have been taken into consideration. Their chemical and isotopic compositions were presented in Table 1.

Table 1 Chemical and isotopic composition of exemplary thermal waters from the Carpathian geothermal province

Name of well	Geothermal system	D (m)	Q (m^3/h)	pH	TDS (g/dm^3)	t_o (°C)	Na^+ (mg/dm^3)	K^+ (mg/dm^3)	Ca^{2+} (mg/dm^3)
Bańska PGP-1	Inner Carpathian, Podhale basin	3242.0	550.0	7.05	2.74	86.0	550.0	55.2	230.5
Lubatówka 12	Outer Carpathians Eastern part	958.0		6.95	19.22	21.2	6620.2	40.3	67.2
Sobniów 29		2240.0		7.53	21.15	25.2	7588.6	138.0	40.0
Jaszczew 32a		2530.0		7.86	15.40	22.1	5310.4	54.0	29.0
Ustroń U-3A	Outer Carpathians, Western part	1753.0	0.6	6.13	118.9	20.8	32190.0	774.0	8836.0

Name of well	Geothermal system	Mg^{2+} (mg/dm^3)	Cl^- (mg/dm^3)	SO_4^{2-} (mg/dm^3)	HCO_3^- (mg/dm^3)	Li^+ (mg/dm^3)	Sr^{2+} (mg/dm^3)	F^- (mg/dm^3)	Br^- (mg/dm^3)	SiO_2 (mg/dm^3)	$\delta^{18}O$ (‰ vs VSMOW)	δ^2H (‰ vs VSMOW)
Bańska PGP-1	Inner Carpathian,	41.8	463.0	920.0	324.0	0.99	6.45	2.49	0.91	66.9	−11.4	−82.3
	Podhale basin	72.8	8300.2	7.0	3980.5	3.86	19.09	0.70	26.21	12.6	1.2	−26.2
Lubatówka 12	Outer Carpathians	9.0	10258.8	154.0	2757.2	2.70	10.36	7.36	66.20	32.0	5.2	−21.2
Sobniów 29	Eastern part											
Jaszczew 32a		10.0	6164.6	272.0	3306.2	2.10	9.89	6.90	105.97	29.0	4.9	−25.0
Ustroń U-3A	Outer Carpathians, Western part	2528.0	73327.0	28.0	128.0	13.40	411.0	0.0	429.0	14.20	−0.6	−19.4

D total depth of the well; t_o outflow temperature; *Q* admissible discharge; *TDS* total dissolved solids

Thermal waters in the eastern part of the Outer Carpathian Geothermal Basin belong to saline of $Cl(HCO_3)$–$HCO_3(Cl)$–Na and Cl–Na hydrochemical types. Their total dissolved solids (TDS) values vary usually from about 15 to 20 g/dm^3 whereas outflow temperatures from 20 to 25 °C. In the eastern part of the Outer Carpathians such thermal waters occur usually in the area of oil fields: folds, local thrust or flexure structures. Thus, they represent both the typical oil-field brines (because they are extracted together with oil as a byproduct, for example waters from wells Jaszczew 32a and Sobniów 29) or the so-called edge oil-field waters (i.e. these which are not in direct contact with oil deposits, like for example, water from the well Lubatówka 12 belonging to Iwonicz Zdrój spa and extracted for production of mineral bathing salts). Both types of waters are additionally enriched in various degrees in such specific compounds as bromine, iodine, boron, barium, strontium, lithium; in the gas phase CH_4 is dominant.

Taking into account other known boreholes with thermal water manifestations both the TDS and outflow temperatures may vary in wide ranges from below 10 g/dm^3 up to about 60 g/dm^3, and from 20 to 84 °C, respectively. Always, the water bearing horizons are connected with flysch sediments.

In the western part of the Outer Carpathians thermal waters are encountered in flysch and subflysch sediments. Due to the fact that the water-bearing capacity of flysch sediments is rather low, the better hydrogeological and thermal conditions are found in underlying strata, mainly Carboniferous sandstones and Devonian carbonates, and sometimes in Carpathian Foreedep Miocene. To date, the fissure water bearing horizons in Devonian carbonates are considered as the most perspective with respect to thermal water occurrence and economic exploitation. One of the examples may be the working thermal well U-3A in Ustroń spa (Table 1). The water extracted from the well is a brine of Cl–Na–Ca type with TDS value of 118.9 g/dm^3 and outflow temperature of 20.8 °C. The largest discharge in the range of 0.65 m^3/h was obtained in the depth interval between about 1,100–1,665 m from Upper and Middle Devonian carbonates. The temperature logging performed at the well revealed the temperature of 38–43 °C in depth interval 1,275–1,350 m, and about 50 °C at the bottom of the well (Sokołowski and Kempkiewicz 2001). Due to the low discharge, the outflow temperature of water is considerably lower than that observed at the bottom of the well.

The Podhale Basin within the Inner Carpathian Geothermal Basin is the best recognized and the most perspective area with respect to thermal water occurrence and utilization. To date about 12 thermal wells have been drilled in this area. The shallowest well, Siwa Woda IG-1, with total depth of 856 m, extracts HCO_3–SO_4–Mg–Na type water of TDS below 0.5 g/dm^3 and temperature 20 °C at the outflow. The deepest well, Bańska IG-1, with total depth of 5,261 m, extracts SO_4–Cl–Na–Ca type water of TDS about 2.6 g/dm^3 and temperature 82 °C at the outflow (Chowaniec et al. 2001). However, the highest outflow temperature of 86 °C were found for SO_4–Cl–Na–Ca type water extracted from the well Bańska PGP-1, total depth is 3,242 m. The majority of thermal wells in this area extract water from Triassic limestones and dolomites and Eocene carbonates.

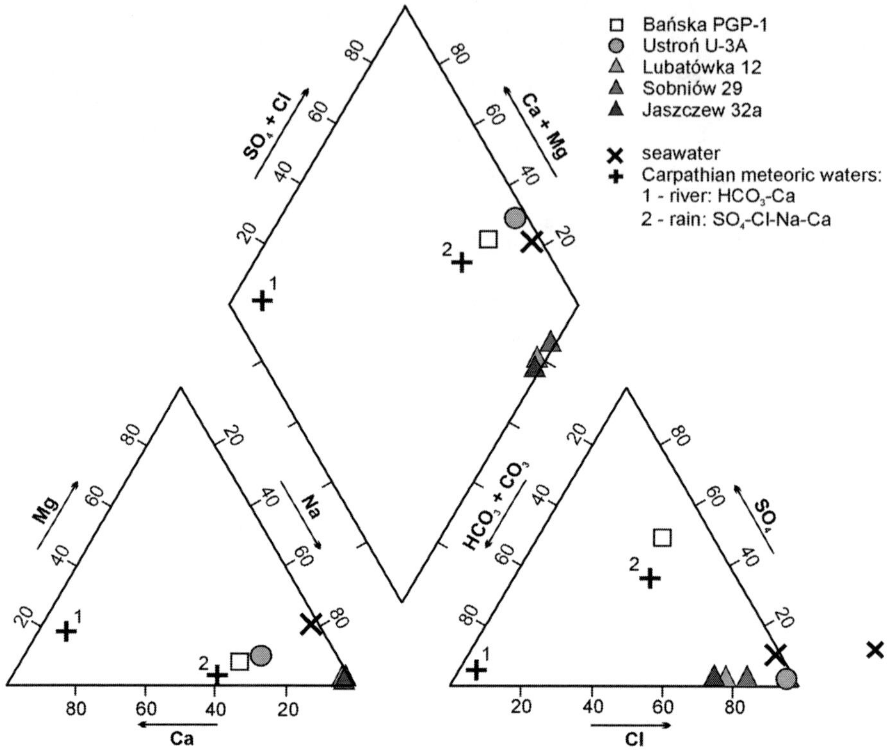

Fig. 5 Graphical presentation of chemical composition of typical thermal waters encountered in the Carpathian geothermal province in Poland. Average chemical composition of seawater and meteoric waters are plotted for reference

The comparison of the thermal waters chemical compositions in the Carpathian Geothermal Province was shown on the Piper diagram (Fig. 5).

As can be seen on the plot, there is a clear chemical difference among waters from the Inner and Outer Carpathians, which also may suggest their different origin. The multi-ionic hydrochemical type, low TDS value and the structure of chemical composition of main compounds similar to rain waters strongly suggest the meteoric origin of the Podhale Basin thermal waters. Moreover, thermal waters extracted from carbonates reveal higher concentrations of Ca^{2+} and Mg^{2+}, which concern both types of waters from Bańska and Ustroń.

Oxygen and hydrogen isotopic composition of thermal waters occurred within the Inner and Outer Carpathians (Fig. 6) corroborate their completely different origin.

The position of the Podhale Basin thermal water directly on the GMWL indicates undoubtedly its meteoric origin. Moreover, the studied water reveals much lighter isotopic composition (i.e. higher concentration of light isotopes ^{16}O and ^{1}H over heavy isotopes ^{18}O and ^{2}H) in comparison to modern meteoric waters (MW)

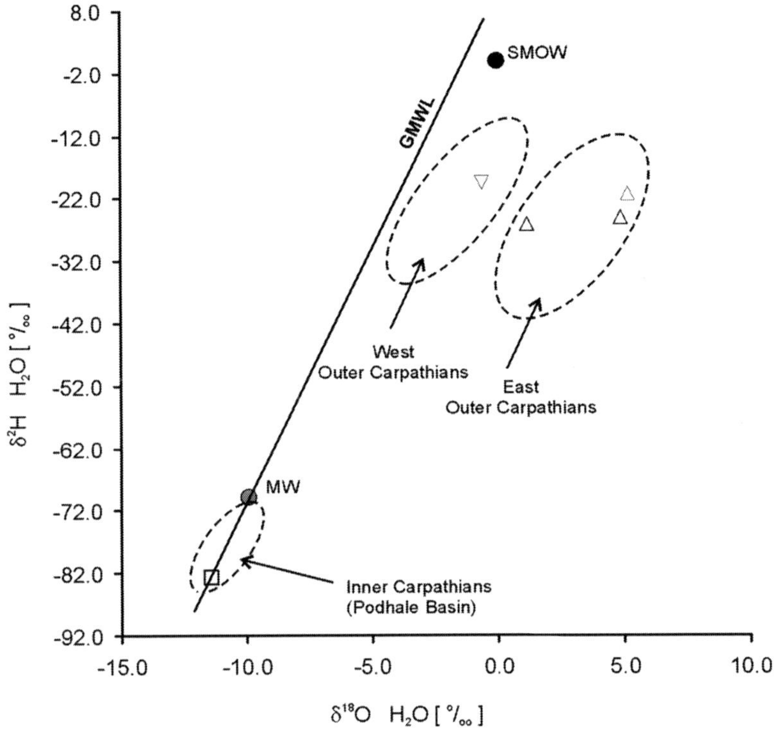

Fig. 6 Oxygen and hydrogen isotopic composition of thermal waters in different parts of the Carpathian geothermal province (*CGP*) in Poland. Global meteoric water line (*GMWL*) after Różański et al. (1993). SMOW and meteoric water (*MW*) are plotted for reference. For more information see the text

represented by the water from Jasiołka River, Outer Carpathians. This strongly suggests that the recharge area of thermal water horizons in Podhale Basin must be located on the higher altitudes, in Tatra Mountains. This agrees with all hydraulic studies, which indicate groundwater flow pathways in the basin directed from north to south, i.e. from Tatra Mountains to Pieniny Klippen Belt.

On the other hand, the isotopic composition of thermal waters in Outer Carpathian indicates their more complex origin. In previous studies directly concerning the mineralized waters of the Central Carpathian Synclinorium it has been shown that the origin of these waters is polygenetic (Dowgiałło 1980; Oszczypko and Zuber 2002; Porowski 2001b, c, 2004a, b, 2006). Probably the majority of them are three component mixtures of: (1) connate waters—those of seawater-like trapped in pores of sediments during sedimentation, and undergone further chemical evolution; (2) dehydration waters, i.e., those originated from dehydration of clay minerals which take place during transformation of mixed-layer illite/smectites from smectite dominated to illite dominated; and (3) meteoric waters which belong to those of modern hydrological cycle or/and palaeoinfiltration ones.

4 Thermal Conditions Derived from the Geothermometric Study

The distribution of the surface heat flow density for the whole Carpathian Geothermal Province is quite variable (Plewa 1994). In the western part, these values are in the range of 41.6–70.0 mW/m^2, between 50.8 and 81.6 mW/m^2 in the middle part together with the Inner Carpathians, and between 27.0 and 55.8 mW/m^2 in the eastern part. Average value of the heat flow density for the whole Province equals 55.4 mW/m^2 (Plewa 1994). Based on the temperature logging in selected boreholes in the area of Polish Carpathians, the calculated geothermal gradient varies from 2.0 °C/100 m to 3.21 °C/100 m, with average value of about 2.35 °C/100 m (Plewa 1994). To evaluate the hydrogeothermal systems in the light of their prospectiveness to further exploration of thermal energy resources, the authors attempted to apply selected isotopic and chemical geothermometers to assess the maximum possible temperatures, which may be encountered in the reservoir (Table 2).

The chemical geothermometry is based on the temperature dependent solubility of various minerals and their phases, for example, silica mineral phases, albite, feldspar, clay minerals, etc. Roughly speaking, the equilibrium between dissolution of primary minerals and precipitation of secondary ones or other phases controlling the concentration of particular cations in water in given temperature indicates, in favorable hydrogeological settings, the dominant temperature in the reservoir (Arnórsson 2000).

In case of thermal waters studied here, the temperatures calculated by quartz and chalcedony geothermometers fit the down-hole and outflow water temperatures of the Podhale and Ustroń hydrogeothermal systems very well (Table 2). The chalcedony geothermometers estimate the outflow temperatures very well, whereas the quartz geothermometer indicates rather bottom-hole temperatures, which correspond to maximum temperatures in the reservoir. Geothermometric calculations made by Kępińska (2001) for thermal waters extracted from the deepest horizons within the Podhale hydrogeothermal system corroborate the very good agreement of temperatures estimated by quartz and chalcedony geothermometers with those measured at the outflow and at the bottom of the wells.

The application of Na–K and Mg–Li geothermometers to thermal water from Podhale Basin give temperatures close to that measured at the outflow, which are however, too low to be considered as the reliable temperatures in the reservoir. In case of hydrogeothermal system in the Ustroń area, both these geothermometers give rather overestimated values of reservoir temperatures. Probably, one of the reasons for such a situation may be quite low water capacity and small well discharge in the Ustroń (which causes fast cooling of a small amount of water during extraction), and the very high salinity of thermal water. Completely out of range are the reservoir temperatures in the Podhale Basin and Ustroń area estimated by oxygen isotope geothermometer in water—sulfate system; the estimated reservoir temperature is much lower than the outflow temperature of water in the

Table 2 The results of application of selected chemical and isotopic geothermometers to the assessment of reservoir temperatures in the studied hydrogeothermal systems

Name of well	Hydrogeothermal System	t_o	t_{bh}	t_{gg}	t_q	t^1_{ch}	t^2_{ch}	t_{Na-K}	t_{Mg-Li}	$t^{18}_{\Delta}O_{M\&R}$
Bańska PGP-1	Inner Carpathians, Podhale Basin	86.0	90	92	116	87	88	79	77	55
Lubatówka 12	Outer Carpathians, Eastern part	21.8		30	48	15	20	18	105	
Sobniów 29		25.2		61	83	52	55	68	126	
Jaszczew 32a		22.1		68	79	47	50	39	116	
Ustroń U-3A	Outer Carpathians, Western part	20.8	55	49	56	24	28	84	91	133

t_o—outflow temperature
t_{bh}—bottom-hole temperature
t_{gg}—temperature at depth estimated by geothermal gradient
t_q—temperature estimated by quartz geothermometer (Fournier 1977)
$t = \frac{1,309}{5.19 - \log S} - 273.15$, where S—$SiO_{2(aq)}$ concentration in ppm
t^1_{ch}—temperature estimated by chalcedony geothermometer (Fournier 1977)
$t = \frac{1,032}{4.69 - \log S} - 273.15$, where S—$SiO_{2(aq)}$ concentration in ppm
t^2_{ch}—temperature estimated by chalcedony geothermometer based on field study (Arnórssson et al. 1983)
$t = \frac{1,112}{4.91 - \log S} - 273.15$, where S—$SiO_{2(aq)}$ concentration in ppm
$t_{Na\text{-}K}$—temperature estimated by Na–K geothermometer (Arnórssson et al. 1983):
$t = \frac{933}{0.993 + \log(\frac{Na}{K})} - 273.15$
$t_{Mg\text{-}Li}$—temperature estimated by Mg–Li geothermometer (Kharaka and Mariner 1987):
$t = \frac{2,200}{\log(\frac{\sqrt{Mg}}{Li}) + 5.47} - 273.15$
$t^{18}_{\Delta}O$ (M&R)—temperature estimated by oxygen isotope geothermometer $\Delta^{18}OSO_4$–H_2O with fractionation factor of Mizutani and Rafter (1969)

well Bańska PGP-1, and is almost 2.5 times higher than the measured bottom-hole temperature in the well Ustroń U-3A. The isotopic composition of sulfates in both waters seems to be typical for marine sulfates of Devonian to Triassic age (Clark and Fritz 1997). However, there is no evidence for oxygen isotopic equilibrium in the sulfate—water system. For the Podhale Basin, it suggests high dynamic of water in the system and/or some other secondary processes, which affect this equilibrium. On the other hand, in the Ustroń area the high salinity of water (i.e. contact with evaporates) and too low temperature in the reservoir might be the reasons that the equilibrium was not achieved.

In the case of thermal waters from the eastern part of the Outer Carpathians the reservoir temperatures estimated by chemical geothermometers are very diverse and variable. Although the chalcedony geothermometer estimate quite well the outflow temperature of water from the well Lubatówka 12, it is not the case for other typical oil-associated waters. In the eastern part of the outer Carpathians, where resources of edge oil-field waters and typical oil-associated waters occur in separate folds, they form practically separate, local hydrogeological systems with characteristic thermal conditions, which are opened for contact with modern meteoric waters in various degrees. That is why also the variable results of geothermometric applications should be expected. Generally, silica and Na–K geothermometers indicate reservoir range of temperatures in the eastern part of the flysch Carpathians equal a few dozen degrees of centigrade (i.e., between 48 and 83 °C for studied exemplary wells). The Mg–Li geothermometer, which is specially recommended for estimation of reservoir temperatures in the areas of oil fields (Kharaka and Mariner 1987), indicate maximum temperatures between 105 and 126 °C. It is difficult to answer the question, whether these temperatures are possible nowadays for groundwater horizons in this part of the Carpathians. Of course, this problem needs further investigation to take into account the complex origin of waters in this area. However, there are some premises for supporting the idea about possibilities of high temperatures (i.e. in the range of slightly more than 100 °C) in deeper parts of the flysch Carpathians. The highest temperatures estimated by Mg–Li geothermometers are connected with the isotopically heaviest waters, i.e. those having the largest admixture of dehydration waters (Porowski 2004b, 2006). Dehydration of mixed layers smectite/illite occurs in each sedimentary basin where shale sediments are subjected to increased temperatures during diagenesis, usually 80–120 °C. According to Perry and Hower (1972) the geothermal gradient of about 2.4 °C/100 m would be enough to initiate dehydration at the depth of about 3,000 m. Moreover, in typical sedimentary basins the generation of oil occurs in temperatures ranging from 50 to 175 °C, with maximum intensity in temperatures of about 90 °C (Hunt 1979). The range of these temperatures in the Central Carpathian Synclinorium coincide with the mineralized water reservoir temperature range between 50 and 130 °C, calculated on the base of the Mg and Li geothermometric ratio (Porowski 2004b, 2006). All these facts seem to support the idea about higher temperatures dominated in the eastern part of the Polish Outer Carpathians in geological past and/or about origin of waters connected with deeper parts of the basin. Nowadays the whole basin is considered to be cooled down, and the reservoir temperatures of mineralized waters calculated with application of Mg–Li geothermometer should be treated tentatively as kind of residual paleotemperatures. However, as was mentioned above in this chapter, in the vicinity of Wiśniowa, Skole Unit, the water with outflow temperature of 84 °C was found in the exploratory borehole, which is a quite optimistic premise for the future explorations.

5 Conclusions

The Polish part of the Carpathian Mountains constitutes the Carpathian Geothermal Province composed of two geologically distinct parts called the Inner Carpathian Geothermal Region and Outer Carpathian Geothermal Region.

The Podhale artesian basin is the main geological structure of thermal water occurrence within the Inner Carpathian Geothermal Region. This area is best recognized and considered as the most perspective geothermal region within the whole Polish Carpathians. The main horizons of thermal waters are connected with deep (from 1,000 to 3,500 m), regional groundwater flow system in Eocene carbonates in the upper part and Middle Triassic limestones and dolomites, sometimes Jurassic sandstones, in the lower part. The geochemical type of thermal waters changes slightly from HCO_3–SO_4–Mg–Na in the upper parts of the aquifer to SO_4–Cl–Na–Ca type in the deepest parts. The outflow temperatures of waters vary from 20 to 86 °C. The multi-ionic hydrochemical type, the structure of chemical composition of main compounds, low TDS values, and first of all oxygen and hydrogen isotopic composition indicate their meteoric origin connected with modern hydrological cycle. The Tatra Mountains are considered to be the recharge area of thermal water aquifers in the Podhale basin. The origin of waters, low TDS, vicinity of recharge area and good hydraulic conditions in the aquifer make this hydrogeothermal system very promising and valuable for thermal water utilization in many aspects. To date about 12 thermal wells have been drilled and thermal waters utilization is growing fast.

The Outer Carpathian Geothermal Region, due to the complex geological and tectonic structure, is poorly recognized with respect to thermal water occurrence, resources and possibilities of their utilization. The thermal waters in this region have been encountered in several exploratory drillings (in a majority connected with oil and gas exploration), and are utilized only in a few places for balneological treatment (for example: Ustroń, Rabka). The main reservoirs of these waters are linked to flysch sediments from Lower Cretaceous to Oligocene; however, in the western part of the region, where thickness of flysch is less, also the subflysch Paleozoic carbonates contain aquifers with hot water. Usually, the perspective horizons with thermal waters occur in depth of about 1,000 m and more. A vast majority of thermal waters in the area of Outer Carpathian is brackish, saline and brine. Especially in the eastern part of the region, within the Central Carpathian Synclinorium, they are associated with oil and gas deposits. Their TDS values vary from below 10 g/dm^3 up to about 60 g/dm^3 in the eastern part and up to about 160 g/dm^3 in the western part; outflow temperatures vary from 20 °C up to about 50 °C, exceptionally reaching 84 °C (for example, the vicinity of Wiśniowa). The geochemical types of thermal waters are variable from HCO_3–Cl–Na through Cl–HCO_3–Na to Cl–Na and Cl–Na–Ca for the most concentrated brines. Nevertheless, the distinction of individual hydogeothermal systems in the area of Outer Carpathians is hardly possible. The recharge areas for thermal waters are still not known, their origin is poligenetic and their resources might be limited. Also the thermal conditions in the area of flysch Carpathians seems to be still poorly recognized, and reveal variable character and locally might be very promising. The geothermometric calculations applied for deep seated waters in this area, although debatable and controversial, indicate some premises for occurrence of prospective thermal water resources. Especially when oil and gas deposits in this area are running down fast, the thermal waters, even low-temperature ones, might occur one of the most important factor of further social and technological development of this area.

References

Arnórssson S, Gunnlaugsson E, Svavarsson H (1983) The chemistry of geothermal waters in Iceland. III. Chemical geothermometry in geothermal investigations. Geochim Cosmochim Acta 47:547–566

Arnórsson S (2000) Isotopic and chemical techniques in geothermal exploration, development and use. IAEA, Vienna

Chowaniec J (1989) Hydrogeologiczne warunki zasilania i przepływu wód podziemnych w utworach trzeciorzędowych Podhala między Zakopanem a Białym Dunajcem. CAG PIG, Warszawa. Doctoral Dissertation, in polish

Chowaniec J (2003) Wody podziemne niecki podhalańskiej. Mod Probl Hydrogeol 11(1):45–51

Chowaniec J (2009) Hydrogeology study of the western part of the polish Carpathians. Biletyn PIG 434:98

Chowanie J, Poprawa D (1985) Selected problems of hydrogeology of Podhale. In: Proceeding reports of the XIII congress of KBGA, Part II. Poland, pp 401–406

Chowaniec J, Poprawa D, Witek K (2001) Occurrence of thermal waters in polish Carpathians. Materiały IV Seminarium, Rola energii geotermalnej w zrównoważonym rozwoju, Bukowina Tatrzańska, 8–10 Październik, 2001, pp 89–114 [in Polish]

Clark ID, Fritz P (1997) Environmental isotopes in hydrogeology. Lewis Publishers, Boca Raton, p 328

Dowgiałło J (1976) Problems of the origin of Cl–HCO3–Na mineral waters of the polish flysch Carpathians. In: Cadek J, Paces T (eds) Proceedings of the international symposium on water-rock interaction, Czechoslovakia 1974. Geological Survey, Prague, pp 12–22

Dowgiałło J (1980) Poligenetic model of the Carpathian chloride waters and some of its consequences. Mod Prob Hydrogeol :275–290, Jachranka [in Polish]

Dowgiałło J, Sławiński A (1979) Remarks on the origin of saline ground waters at Rabka (West Carpathians). In: Słowańska B, Pakulska Z (eds), Hydrogeochemistry of mineralized waters. Proceedings of the IAH conference of Cieplice Spa (Poland), 1978. Instytut Geologiczny, Warszawa, pp 195–203

Dowgiałło J, Leśniak PM (1980) The origin of chloride waters in the polish flysch Carpathians. In: Hitchon B (ed) Proceedings of the 3rd international symposium on water-rock interaction, Canada, pp 20–23

Fournier RO (1977) Chemical geothermometers and mixing models for geothermal systems. Geothermics 5:41–50

Hunt JM (1979) Petroleum geochemistry and geology. W. H Freeman and Company, San Francisco, p 617

Kępińska B (1995) Temperature of the main aquifer of the Podhale geothermal system. Technika Poszukiwan Geologicznych 6:3–14 [in Polish, English abstract]

Kępińska B (2001) Hydrothermal and thermal conditions within the Podhale geothermal system in the area of Biały Dunajec PAN-1 well. Studia, Rozprawy, Monografie 93:142 [in Polish; English summary]

Kharaka YK, Mariner RH (1987) Chemical geothermometers and their application to formation waters from sedimentary basins. In: Naeser ND, McCulloh TH (eds) Thermal history of sedimentary basins. Methods and case histories. Springer, New York, pp 99–118

Książkiewicz M (1972) Geological structure of Poland. Vol. IV: tectonics, Part 3: Carpathian Mts. Wydawnictwa Geologiczne, Warszawa 121–203 [in Polish]

Książkiewicz M (1975) Bathymetry of the carpathian flysch basin. Acta Geol Pol 25(3):309–367

Kuśmierek J (ed) (1995) Evolution and hydrocarbon potential of the polish Carpathians. Prace Geologiczne 138:9–88 [in Polish; English summary]

Leśniak PM (1980) The origin of chloride waters at Wysowa, West Carpathians. Chemical and isotopic approach. Acta Geol Pol 30(4):519–550

Mizutani Y, Rafter TA (1969) Oxygen isotopic composition of sulfates—3Oxygen Isotopic Fractionation in the Bisulphate Ion-Water System. NZ J Sci 12:54–59

Nescieruk P, Paul Z, Ryłko W, Szymakowska F, Wójcik A, Żytko K (1992) Geological map of Poland to the scale of 1:200,000, Jasło Sheet: B—map without quaternary sediments. Wydawnictwo Kartograficzne Polskiej Agencji Ekologicznej S.A. [in Polish]
Oszczypko N, Zuber A (2002) Geological and isotopic evidence of diagenetic waters in the polish flysch Carpathians. Geol Carpath 53:257–268
Perry EA Jr, Hower J (1972) Late-stage dehydration in deeply buried pelitic sediments. Am Assoc Pet Geol Bull 56:2013–2021
Plewa S (1994) Distribution of geothermal parameters on the territory of Poland. CPPGSMiE PAN Press, Kraków, p 138 [in Polish; English summary]
Porowski A (2001a) Temporal and spatial variability in chemical composition of mineralized waters from the Iwonicz Anticline, SE Poland. Przegląd Geologiczny 4:317–325 [in Polish; English abstract]
Porowski A (2001b) Chemical composition and origin of ionic ratios in the Iwonicz anticline mineralized groundwaters (polish outer Carpathians). Ann Soc Geol Pol 71:125–134
Porowski A (2001c) Chemical indicators of the origin of mineralized waters in the Central Carpathian Synclinorium. In: BocheńskaT, Staśko S, Wąsik M (eds), Modern problems of hydrogeology, vol X(1). Wrocław, pp 397–404
Porowski A (2004a) Isotopic evidence of the origin of mineralized waters from the Central Carpathian Synclinorium, SE Poland. Environ Geol 46:661–669
Porowski A (2004b) Origin of mineralized waters in the Central Carpathian Synclinorium. ING PAN. Doctoral Thesis, p 195 [in polish; unpublished]
Porowski A (2006) Origin of mineralized waters in the Central Carpathian Synclinorium, SE Poland. Stud Geol Pol 125:67
Różański K, Araguas-Araguas L, Gonfiantini R (1993) Isotopic patterns in modern global precipitation. In: Stewart PK, Lohmann KC, McKenzie J, Savin S (eds) Climate change in continental isotopic records. Geophysical Monograph 76:1–36
Ryłko W, Paul Z (1998) Geological map of Poland to the scale of 1: 200000, Cieszyn sheet. Polish Geological Institute Press, Warszawa [in Polish]
Sokołowski J (1985) Occurrence of thermal waters in the Podhale basin. In: Proceedings of conference, Ocena możliwości eksploatacji wód termalnych w niecce podhalańskiej, Zakopane 21 Aug 1985. AGH 25–46
Sokołowski J, Kempkiewicz K (2001) The assessment of geothermal resources in the area of Ustroń municipality and their utilization. In: Proceedings of international scientific conference geothermal energy in underground mines, 21–23 Nov 2001, Ustroń, Poland, pp 107–114
Ślączka A (1977) Remarks on the geological structure of the Krosno Province. In: Ślączka A (ed) Guide to XLIX meeting of polish geological society. Wydawnictwa Geologiczne, Warszawa, pp 7–32 [in Polish]
Zuber A, Grabczak J (1985) The origin of some mineral groundwaters of Poland in the light of up-to-date isotopic studies. Modern problems of hydrogeology. In: Proceedings of the 3rd national symposium, Wydawnictwo Akademii Górniczo—Hutniczej, Kraków, pp 135–148 [in Polish]
Zuber A, Grabczak J (1986) Determining the origin of mineral waters in southern Poland with the aid of isotope methods. Freib Forsch 417:116–126
Zuber A, Grabczak J (1987) On the origin of chloride waters in the flysh Carpathians—continuation of the discussion. Przegląd Geologiczny 7:366–372 [in Polish; English abstract]

Helwan Springs

Ibrahim El Shamy

Abstract The springs in Helwan, southern suburb of Cairo, were known since the Pharaonic era. Also, the curative features of the Ain Helwan spring waters were first discovered by the ancient Egyptians. Nowadays the area of occurrence of Helwan springs is considerably polluted and turned into barren wasteland. The chapter shows the actual chemical composition of waters from two types of Helwan springs, namely the mineral springs and the sulfur springs containing the H_2S. The hydrogeological conditions of the occurrence of the springs are presented in the light of their vulnerability to antropopression.

Keywords Groundwater contamination · Mineral and thermal waters · Helwan springs · Egypt

1 Introduction

Helwan district is one of the distinguished recreation areas, not only in Egypt but all over the world. It has a significant therapeutic effect on several diseases, due to existing sulfur and mineral springs and the dry air. Nowadays, Helwan is surrounded by many factories, which along with the growth of population pollute the groundwater. This pollution contaminates the existing springs that issue from deep unpolluted aquifers. This means that the water of Helwan springs is of good quality until it is encroached by polluted water within the upper 50 m.

Consequently, Helwan groundwater and springs suffer from poor quality management. This chapter presents general characterization of mineral and thermal waters of Helwan with regard to the hydrogeological and geological settings in the area.

I. El Shamy (✉)
Department of Hydrogeology, Helwan University, Greater Cairo, Helwan, Egypt
e-mail: adamp@twarda.pan.pl

W. Balderer et al. (eds.), *Thermal and Mineral Waters*, Environmental Earth Sciences,
DOI: 10.1007/978-3-642-28824-1_8, © Springer-Verlag Berlin Heidelberg 2014

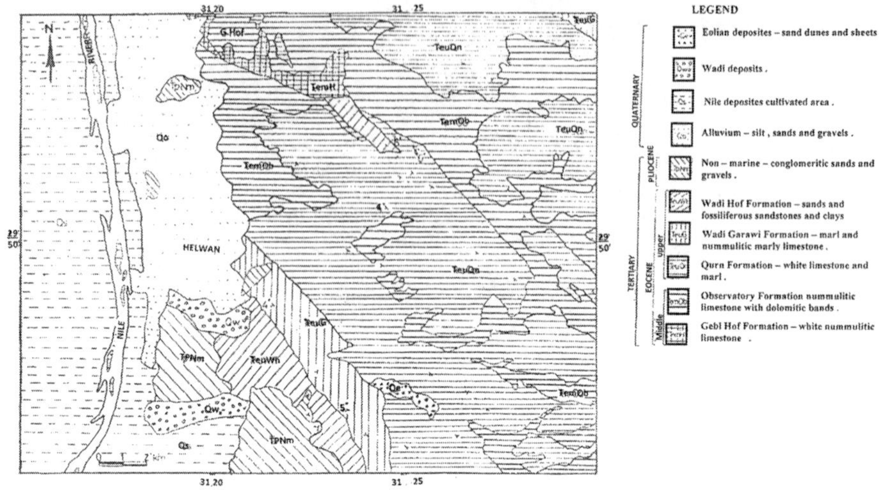

Fig. 1 Geological map of the study area, compiled from the Geological Survey of Egypt (1983) and Conoco Corel (1987)

2 Geology and Occurrence of Mineral Water

The occurrence of Helwan springs depends on the prevailing geologic conditions.

First of all, the area with spring emanations is occupied by thick sedimentary succession mostly of Eocene and Cretaceous carbonates which are underlain by Lower Cretaceous sandstones, so-called "Nubian Sandstones" (Fig. 1). Frequently, such rocks contain some evaporates, which in the occurrence of the sulfur-reducing bacteria change natural sulfates (SO_4) in water to hydrogen sulfide. Hydrogen sulfide often is present in wells drilled in shale or sandstone.

The occurrence of Helwan springs is controlled by the existing fault system, mainly of NW–SE direction. Such faults are rejuvenated by the frequent occurrence of earthquakes (Fig. 2).

Springs of Helwan can be classified into two groups, namely: (i) the sulfur springs, and (ii) mineral springs. The classification depends on the occurrence of H_2S, water salinity and water temperature. Sulfur springs are characterized by the occurrence of H_2S, relatively high salinity and temperature. Mineral springs, on the other hand, are characterized by the absence of H_2S, less salinity and lower temperature.

In the light of geological conditions the sulfur springs are presumably issued from deeper horizons than the water of the mineral springs or they may follow through another passage and at a different rate (Abd El Dayem 1971).

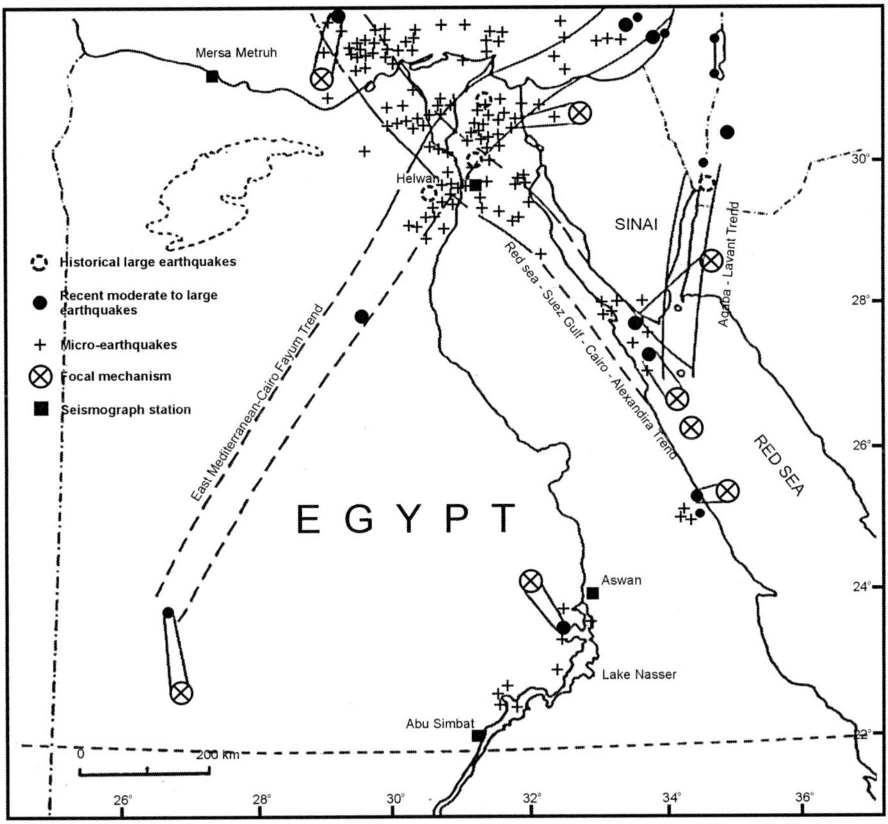

Fig. 2 Sketch showing that the area of eastern Cairo is vulnerable to seismic activity reaching 4–5 on Richter scale (Kebesy 1990). Faults in the area are the suspected cause

3 Chemical Characteristics of Spring Water in the Study Area

Chemical data for the studied spring water are presented in Table 1.

As can be seen in Table 1, the temperature and water salinity of the sulfur springs are relatively higher than that of mineral springs.

HCO_3 ions exist with very small values. The maximum (8.5 and 6.3 epm %) were noted for the main sulfur spring (Abd El Dayem 1971).

Generally, the salinity of waters is higher than 5,000 ppm except for Mineral Spring (number 6, Table 1) and Street No. 16 Spring (number 10, Table 1; Sourour 1968; Abdou 1994).

Helwan Sulfur springs (Nos. 1, 3, 4, 5, Table 1):

Concentrations of Ca are less than Mg except for No. 5, (Sourour 1968). The prevailing type of cations arrangement is: Na > Mg > Ca, except for No. 5

Table 1 Chemical composition of Helwan spring water

No.	Spring name	Date of sampling	pH	Temp. (°C)	TDS (mg/dm^3)	Cations (epm %)				Anions (epm %)			Author
						Ca	Mg	Na	K	HCO_3	SO_4	Cl	
1	H. Sulfur	1964	7.9	31.6	6,208	14.9	16.15	68.1	0.76	3.7	15.8		El Ramly (1969)
2	H. New S.	1964	7.5	26.4	5,080	17.3	20.9	61.3	0.56	2.1	18.8	79.05	El Ramly (1969)
3	H. Sulfur Main	Jan 1967	7.6	32.0	6,220	13.5	13.8	72.4	0.7	8.5	15.3	76.1	Abd El Dayem (1971)
4	H. Sulfur Main	Oct 1967	7.8	31.8	6,000	15.6	19.8	64.6	1.0	6.3	16.5	77.1	Abd El Dayem (1971)
5	H. Sulfur	1968	7.9	32.0	5,800	16.6	16.0	66.5		5.6	17.5	76.9	Sourour (1968)
6	H. Mineral S.	1968	8.0	24.0	4,500	18.3	19.1	61.9		4.1	20.9	75.0	Sourour (1968)
7	Str. no 5	June 21, 1992	6.9	31.5	5,626	25.1	13.7	60.6	0.6	3.7	19.2	77.0	Abdou (1994)
8	Str. no 5	Oct 30, 1992	7.1	29.0	5,364	13.6	16.6	68.8	0.95	3.84	15.2	80.9	Abdou (1994)
9	Str. no 5	Dec 10, 1992	7.7		5,520	17.7	11.8	69.7	0.75	4.52	34.64	60.8	Abdou (1994)
10	Str. No 16	Apr 26, 1992	7.5	32.0	4,545	16.5	16.5	66.1	0.89	3.46	20.1	76.4	Abdou (1994)
11	El-Set Khadra	Oct 23, 1992	7.0	29.5	6,092	17.1	19.9	62.1	0.91	3.56	27.2	69.2	Abdou (1994)
12	El-Set Khadra	Dec 10, 1993	7.3		5,658	15.4	14.5	69.3	0.74	3.70	32.1	64.2	Abdou (1994)
13	El-Set Khadra	1996	7.9		5,190	16.0	19.0	65.0		4.0	36.0	60.0	Tawfik (1999)
14	El-Set Khadra	1997	7.2		5,410	18.0	20.0	62.0		4.0	37.0	59.0	Tawfik (1999)

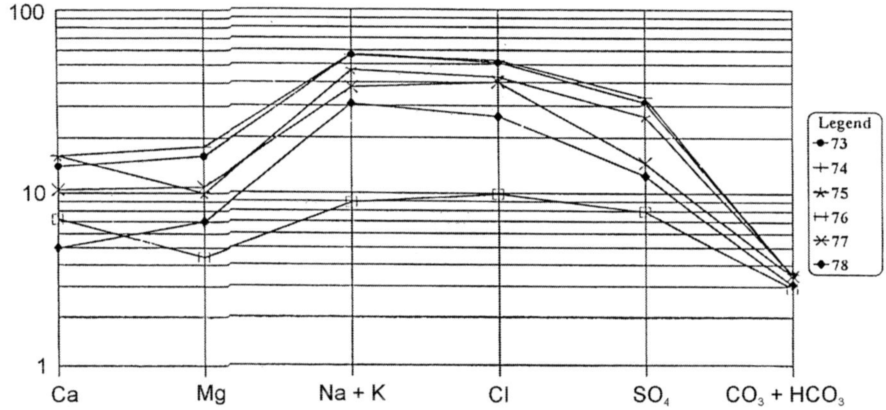

Fig. 3 Chemical composition of Helwan spring's water presented in form of Schoeller diagram (after Tawfik 1999). Numbers refer to following springs: *73* El Set Khadra 1, *74* El Set Khadra 2, *75* El Ezba El Bahriya, *76* Street 5, *77* Mineral Spring 1, *78* Mineral Spring 2

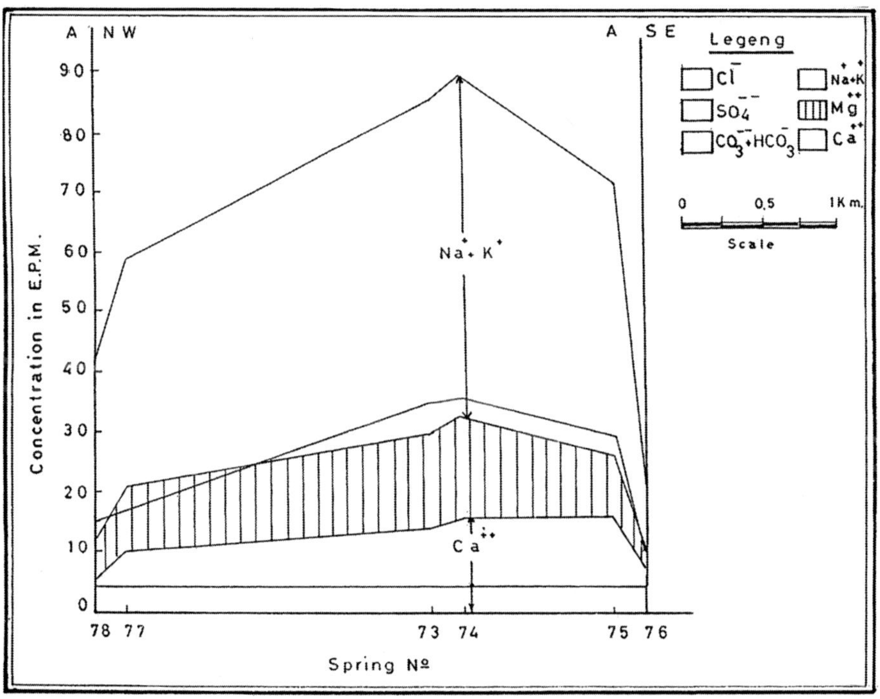

Fig. 4 Hydrochemical profile of natural spring water from NW to SE

(Sourour 1968). Anions arrangement is the same for all springs, namely: Cl > SO_4 > HCO_3. Salinity varies between 6,208 and 5,800 ppm.

Set Khadra springs (Nos. 11–14, Table 1):

Salinity of water varies between 6092 and 5190 ppm, indicating brackish-type water. Concentration of Na^+ is higher than Cl^-, suggesting the meteoric origin for the spring water. Ca^{2+} concentration is less than Mg^{2+}, except for spring No. 12 (Abdou 1994). Concentration of SO_4^{2-} ion is higher for water of this spring than all other springs. Its value ranges between 37.0 and 27.2 epm %. In other springs (except spring No. 9) content of SO_4^{2-} is <21 epm %.

Street 5 Springs (Nos. 7–9, Table 1):

Concentrations of Na^+ ion are less than Cl^- ion, except for spring No. 9 (Abdou 1994). Concentrations of SO_4^{2-} ion are less than 20 epm % for spring Nos. 7 and 8, while it is 34.64 epm % for spring No. 9 (or 35 epm %; after Tawfik 1999).

The relationship among concentrations of cations and anions in the studied spring waters are presented by Schoeller diagrams on Fig. 3. The hydrochemical profile of spring waters from NW to SE is presented on Fig. 4.

According to Tawfik (1999), most of the minor and trace constituents of Helwan spring waters as, B, Fe, Pb, Mn, Cd, Cr, Co and Ni, are within the permissible range of contamination. On the other hand, El Set Khadra spring No. 2 is considered to be polluted by nitrates (NO_3) where its concentration is about 54.97 mg/l, and Street 5 spring—by phosphate (PO_4), where its content reaches a value of about 1.72 mg/l.

References

Abd El Dayem AA (1971) Hydrogeological studies of springs in the area to the east of Cairo. M.Sc. thesis, Faculty of Science, Ain Shams University

Abdou IIM (1994) Geology and hydrogeology of the Helwan area, Egypt. M.Sc. thesis, Faculty of Science, Menoufia University

Conoco Coral and Egyptian General Petroleum Corporation (1987) Geological map of Egypt, sheet no. 36, SW Beni Suef, scale 1:500,000. General Chairman Map project

El Ramly IM (1969) Recent review of investigations of the thermal and mineral springs in the A.A.R. In: 23rd international geological congress, vol 19, pp 201–213

Geological Survey of Egypt (1983) Geological map of greater Cairo area, Egypt, scale 1:100000. Geological Survey Egypt, Cairo

Kebesy RM (1990) Seismicity. In: Said R (ed) Geology of Egypt. Balkema, pp 51–59

Sourour AM (1968) Geological, geophysical and geochemical studies for mineral water in the area east of Cairo till Tebbin. M.Sc. thesis, Faculty of Science, Ain Shams University

Tawfik NS (1999) Pollution of natural water at Helwan area and its vicinities. M.Sc., Faculty of Science, Helwan University

Occurrence of High Bicarbonate Groundwater in Victoria, Australia

Andrew Shugg

Abstract Determination and collection of major ion chemistry for Victorian groundwater was initiated by the Geological Survey in about 1855. The Survey collection is the major source of groundwater information, although in the last two decades water authorities and industry bodies have gathered large amounts of information. The data set compiled for examination of the occurrence of the bicarbonate ion is a synthesis of data from mainly government sources and represents samples mainly from farm bores, government investigation holes and petroleum exploration drilling. Overall bicarbonate is the second most abundant anion in Victorian groundwater with a salinity above 300 mg/L Total Dissolved Salts (TDS). A concentration of 800 mg/L bicarbonate was selected to define the class of high bicarbonate alkalinity groundwater. The occurrence of high bicarbonate groundwater in Victoria has been examined in relation to hydrogeological setting. The chemical evolution to a high bicarbonate groundwater type is observed in a number of different geological systems. Different facies may be identified based on the pH and the predominance of anions and cations including iron. Within each aquifer type, high bicarbonate groundwater can occur within the flow system: in mixing and flushing zones, in hypersaline environments, and spectacularly associated with waste disposal plumes. Geological associations include: fluviatile and alluvial aquifers, extensive aquifers of the coastal sedimentary basins, and the low flow fissure networks in bedrock of the gently undulating ranges of Central Victoria. In the sedimentary coastal basins a common association is with aquifers containing accessory concentrations of carbonate, carbonaceous matter and pyrite. The facies in these aquifers can develop over a short distance and may occur in strata enveloped within other strata containing significantly different water quality. In Central Victoria, Lower Palaeozoic rocks frequently exhibit an association between the occurrence of high bicarbonate groundwater and labile interbedded, lithic and arkosic sandstones with pyritic carbonaceous shales that exhibit deep

A. Shugg (✉)
Federation University, University Drive, Mt Helen, Ballarat, VIC 3350, Australia
e-mail: ashugg@ballarat.edu.au

W. Balderer et al. (eds.), *Thermal and Mineral Waters*, Environmental Earth Sciences, DOI: 10.1007/978-3-642-28824-1_9, © Springer-Verlag Berlin Heidelberg 2014

kaolinised weathering profiles. In deep bedrock mining activities of nineteenth century in Central Victoria dewatering of large volumes of the rock mass took place. Gradually this resulted in a freshening of the groundwater extracted and suggested a localized increase in flow and recharge to the bedrock. A century later and more than 70 years after the dewatering operations stopped, mineral exploration bores encountered groundwater on the same gold fields with increased sulphate and bicarbonate concentrations. The high bicarbonate concentrations had been re-established as a consequence of reaction with sulphide and carbonate minerals.

Keywords Hydrogeology · Mineral water · Chemical composition · Geochemical evolution · High bicarbonate water · Australia

1 Introduction

The State of Victoria is situated on the southern margin of the Australian mainland (Fig. 1). The climate varies across the state from high rainfall alpine in the eastern portion of the State to semi-arid in the northwest. The topography of the State is represented by low rounded mountain ranges that reach elevations of around 2,000 m in the east of the State to the low elevation dune systems of the arid regions of the northwest of the State. As a measure of aridity, less than 30 % of rainfall becomes runoff in Victorian river systems. Seasonal daily maximum temperatures range in the summer from the mid-20s C in the Alpine regions to the mid-40s C in the drier hotter northwest. Winter temperatures are usually between 15–25 °C cooler than in the summer.

2 Defining High Bicarbonate Groundwater

Defining the limits for "high" can be arbitrary, statistical or follow the course of previous literature. On balance, a concentration of 800 mg/L as bicarbonate (HCO_3^-) has been adopted. This fulfils the various criteria of acceptance and statistically only around 5 % of Victorian groundwater samples have a bicarbonate concentration above this level.

3 Data of Victorian Groundwater Chemistry

Information contained in Victorian public groundwater databases has been compiled to examine the statistical distribution and abundance of major ions (Fig. 2). The Victorian groundwater databases contain records for more than 130,000 bores.

Fig. 1 Victoria situated in South Eastern Australia

However, less than 15 % of the bores have a full suite of major ion determinations that includes bicarbonate. The earliest extent high bicarbonate water analysis was made by Maund (1855).

4 Occurrence in Victoria

4.1 Major Ions Distribution

The cumulative frequency curves for the major ions have been plotted on Fig. 2, most of the ions exhibit similar distribution curves but with different statistics. In the universal sample set, waters with a salinity of more than 300 mg/L TDS usually have bicarbonate as the second most abundant anion in groundwater. Although in some high bicarbonate groundwaters, it may be the dominant anion. Statistically, around 15,000 major ion analyses are available for bicarbonate ion concentration examination. Around 20 % of the groundwater samples have more than 600 mg/L and around 6 % have greater than 800 mg/L bicarbonate. A median value for bicarbonate is 269 and the modal value is 360 mg/L (bicarbonate

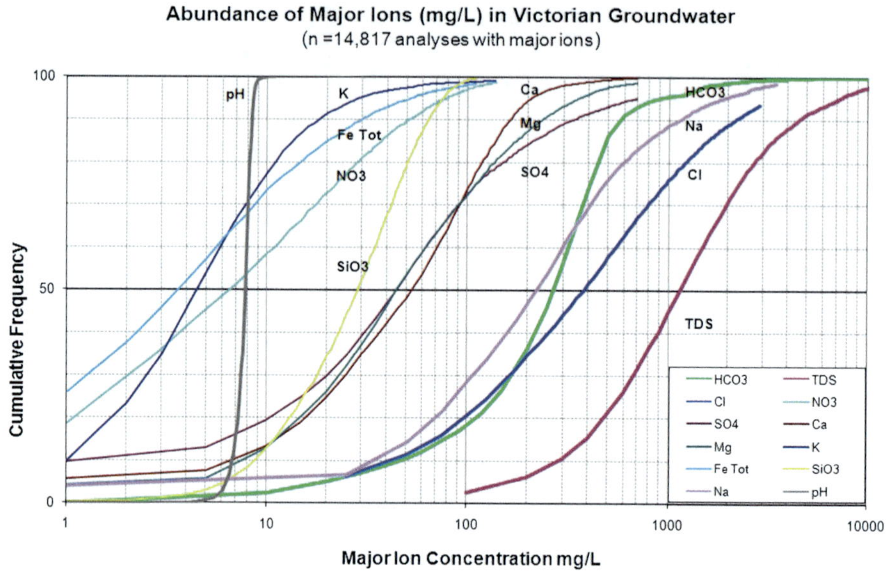

Fig. 2 Major ion frequency distributions for Victorian groundwater

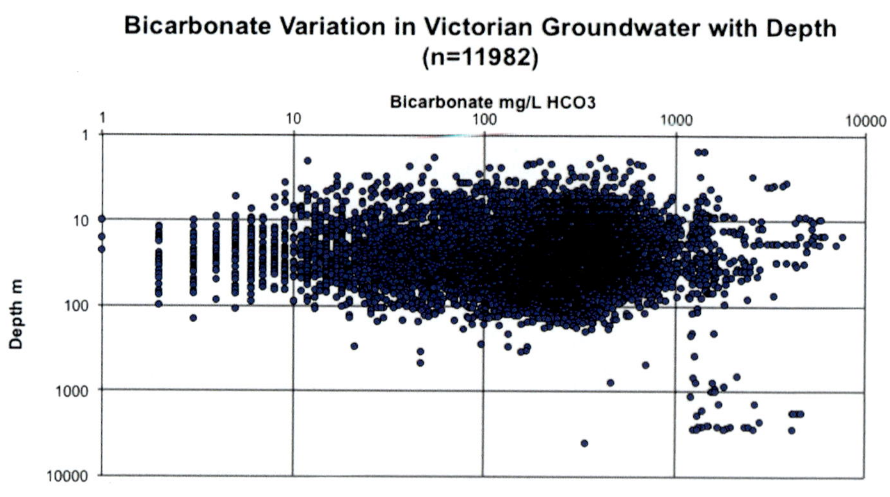

Fig. 3 Bicarbonate depth relationship for Victorian groundwater

as HCO_3^-). The skew of the distribution is illustrated in Fig. 3, and it has a tail towards the high values, which is similar to that of other major cations and anions found in groundwater.

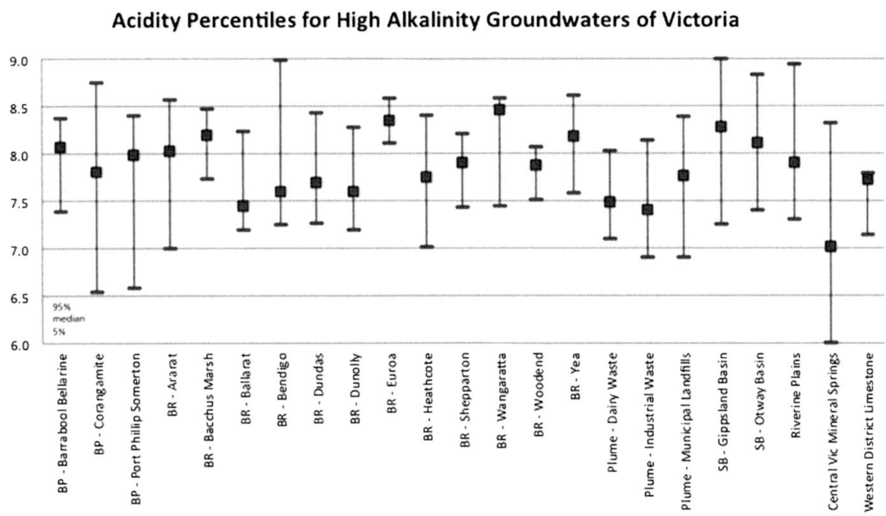

Fig. 4 Range of pH encountered in high bicarbonate groundwater in Victoria

4.2 Bicarbonate and pH

High bicarbonate groundwater may have a wide range of pH's and the percentiles for sites around Victoria have been plotted on Fig. 4. Within the group of high bicarbonate groundwater are those that are naturally effervescent, the cold carbonated mineral waters. Generally, waters with a pH median value of less than 7.2 correspond with those that are effervescent. Many high bicarbonate groundwaters are dominated by the sodium and bicarbonate ions such as the deep sedimentary basin bores, and these have median pH values of above 8; a dilute sodium bicarbonate solution will have a pH of 8.4. Gassy waters possess lower pH values, on Fig. 4, these are identified as the Central Victorian waters. Some gassy waters with a higher pH occur in the coastal sedimentary basins, landfill plumes and dairy waste disposal plumes and the water can also be vigorously effervescent.

4.3 Depth Relationships

Many other criteria may be used to characterise the range and trends in composition of the high bicarbonate waters. Median values of dominant cations and the sodium bicarbonate ratio have been used to illustrate the characterisation. A clustering of the bicarbonate concentrations in groundwaters from a wide range of depths is illustrated in Fig. 3. Deep bore outliers on this diagram illustrate the sampling bias of the database towards shallow bores. The deep high values correspond with off shore oil field formation water, but otherwise fall within the

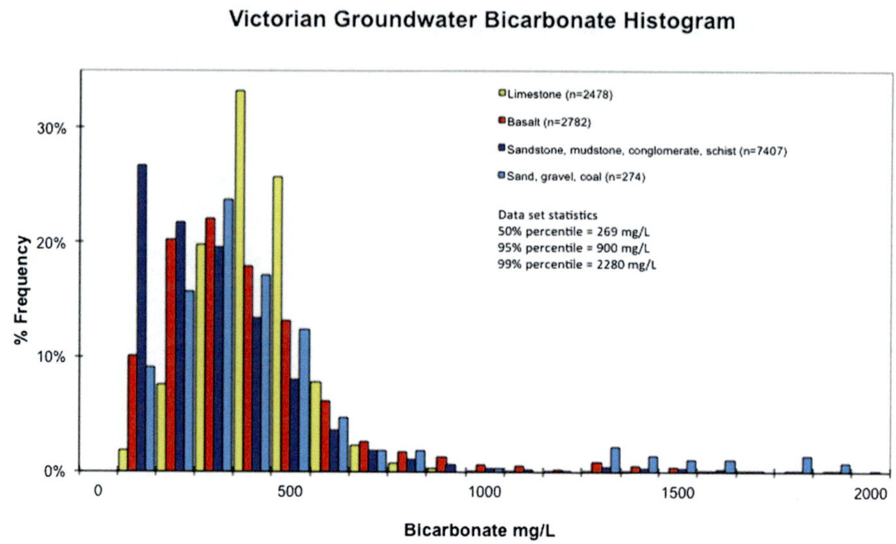

Fig. 5 Bicarbonate frequency distributions for different lithologies

massive cluster values. Another means of differentiating the high bicarbonate groundwater is by cation speciation, reflecting the dominance of particular cation exchange processes and the geochemistry of the aquifer material.

4.4 Lithological Associations

Correlation with different lithologies illustrates slightly different modal values, and the existence of outlier values. These are shown on Fig. 5. All lithological types possess skewed distributions. Carbonaceous sands and sandstones exhibit the longer tail of outliers with frequent values of bicarbonate above 1,000 mg/L. Implicit in the lithological correlations are linked rock water reactions.

5 Geological Distribution

High bicarbonate water as a hydrogeochemical facies can be recognised to exist in several different aquifer and lithological settings across the State. Several categories of occurrence may be associated with either point discharge, a spring, or can be found at shallow or relatively great depths within local or extensive groundwater flow systems. Generally, the higher concentrations are found in shallow bores in Lower Palaeozoic arkosic sandstones or in the deep sedimentary basin bores associated within carbonaceous sands, notwithstanding, mixing of waters particularly those associated with underground waste disposal (Fig. 3).

In Victoria, high bicarbonate groundwater is categorized into four hydrogeological settings, these are:

1. Highland—Bedrock Association—Low flux perennial flow systems in fissured bedrock with waters moving through redox zoning in thick weathering profiles developed in lithic or arkosic sandstones and pyritic carbonaceous shales;
2. Riverine Plains—in transition zones in shallow alluvial aquifers near mixing or flushing zones close to the highland margins;
3. Coastal Margin Sedimentary Basins—in aquifer formations containing accessory carbonate, labile carbonaceous sandy sediments, or impure limestone;
4. Anthropomorphic—Waste disposal and landfill plumes co-exhibiting redox and chromatographic zoning.

Within each setting the predominant lithological settings are presented in Table 1.

5.1 Highland Bedrock Associations

The bedrock occurrences are associated with fractured rocks of Cambrian to Devonian age. The high bicarbonate groundwater is associated mainly with shale sandstone sequences possessing arkosic or lithic sandstone with pyrite and carbonate (ankerite) as accessary minerals. The sandstones possess feldspar and labile lithic particles. Below the water table towards the base of the weathering profiles these sediments usually contain 0.5–2 % carbonate and the interbedded shales possess similar levels of carbon in carbonaceous material. The high bicarbonate waters of the bedrock flow systems contain several water types and include: saline, "fresh"—low chloride, low chloride—low sulphate, carbonated, and a high sulphate facies. The carbonated high bicarbonate groundwater may be differentiated by cation dominance. The dominant chemical processes in the closed systems include: carbonate dissolution, ion exchange and clay mineral reactions which result in the formation of high sodium or sodium—magnesium bicarbonate waters.

The hydraulic capacity of these systems was tested during the gold field dewatering operations between 1860 and 1940. Water samples collected during and after the mine dewatering episodes on the Central Victorian gold fields provides an insight into the geochemical processes and the recovery of chemical characteristics. Dewatering depths varied from greater than 1,600 m at Bendigo, 600 m at Maldon, 540 m at Clunes and 450 m at Ballarat.

Recent data is available in the Mines Department Reports, from exploration company investigations (Australian Groundwater Consultants (AGC), 1983, Bendigo goldfield dewatering, preliminary assessment, open file document for Environment Effects Study (EES) Bendigo Gold, AGC "unpublished"; Forbes and Showers 1988) and from open-file environmental compliance documents. After decades of quiescence, the post recovery waters across the gold fields shows a pattern of water level recovery and the groundwaters show an increase in the

Table 1 Lithological associations of high bicarbonate groundwater in Victoria

Bedrock aquifer systems	Bore depths to	Lithology
Ararat	30	Cambrian to Ordovician (undifferentiated) bedrock sandstone—shale, grit and conglomerate
Bacchus Marsh	50	Some in shallow basalt and others in the underlying or outcropping Ordovician bedrock sandstone and grit
Ballarat	50	Ordovician bedrock, granite and schist and overlying basalt
Bendigo	25	Ordovician bedrock, mudstone, sandstone, grit and conglomerate
Central VIC mineral springs	23	Ordovician bedrock
Dundas	22	Basalts, mudstone and sandstone
Dunolly	21	Ordovician schists, sandstones, grits, conglomerate and mudstone
Euroa	5	Quaternary Shepparton Formation and Devonian Extrusive Acid Volcanics—Violet Town Volcanics
Heathcote	30	Cambrian to Devonian mudstone, sandstone, grit, conglomerate
Shepparton	20	Cambrian–Ordovician sandstones, grits and mudstone
Wangaratta	15	Riverine Plains sediments; sand, mudstone and conglomerate
Woodend	32	Ordovician and through quaternary basalt overlying the Ordovician
Yea	26	Silurian–Devonian limestone, sandstone, grit and mudstone
Anthropomorphic		
Dairy waste plume	50	Miocene limestone, and milk whey plume, fermenting milk products and dairy factory waste
Industrial waste landfill's	35	Various, unconsolidated Tertiary sands, basalts, silts and carbonaceous material
Municipal landfill surrounds	24	Various, unconsolidated Tertiary sands, basalts, silts and carbonaceous material
Riverine plains—highland front		
Riverine plains	22	Tertiary sediments; limestones (MGL), sands and silts
Basalt and coastal plains		
Barrabool Bellarine	70	Shallow bores in quaternary sands, gravels and sandstone. Deep bores 200–300 m deep in Eocene conglomerate—mudstone
Corangamite	30	Upper Tertiary basalts and Miocene–Pleistocene limestone aquifers
Port Phillip Somerton	36	Tertiary basalt or underlying Silurian–Devonian mudstone
Coastal sedimentary basins		
Gippsland basin	1,930	Lower Tertiary and upper Cretaceous aquifers associated with carbonaceous sediments, coals and petroleum reservoirs of Bass Strait
Otway basin	1,370	Lower Tertiary upper Cretaceous aquifers associated with carbonaceous sediments, coals and petroleum reservoirs and limestone aquifers
Western district limestone	22	Miocene limestone

Table 2 Net changes in concentration between the dewatered and post recovery periods for several deep mining districts of Central Victoria

Period of observation	Ballarat 1910–1938	Bendigo, New Chum 1906–1987	Maldon mine water 1867–1938
TDS	9,868	5,771	1,102
HCO_3	679	2,408	539
SO_4	967	900	75
Mg	501	611	80

dissolved salts and the bicarbonate levels, while some of the gold fields show a marked increase in sulphate concentrations relating to the oxidation of sulphide material in the rock mass (Table 2). The importance of these observations lies specifically in the identification of the geochemical process and implicitly in the recovery of the antecedent—pre-mining water chemistry.

5.2 Coastal Basins

High bicarbonate groundwater develops in the shallow basalt aquifers and limestone aquifers and in the deep aquifers of the sedimentary sequences of the coastal plains. These associations with minor variations are most prevalent in the Lower Tertiary and Upper Cretaceous Dilwyn and Waarre Formations of the Otway Basin. Similar high bicarbonate waters develop in the offshore Latrobe Formation of the Gippsland Basin and in the Eastern View Formation coal measures onshore at Anglesea of the Bass Basin. In the sedimentary basin waters, sodium is usually the dominant cation.

5.3 Riverine Plains

Shallow fluviatile aquifers of the extensive alluvial plains also develop high bicarbonate groundwater. The most pronounced occurrence is in close proximity to the Highland margin. The spatial distribution in bores reflects a bias in drilling targets, which seek to find fresh water. Saline areas are not drilled. Thus, data distribution reliability for this area retains some uncertainty. The alluvial aquifers become thinner near the Highland margin and fresh water moving from the uplands mixes and displaces more saline water down the flow systems.

5.4 Anthropomorphic Sources

High bicarbonate groundwater occurs associated with plumes emanating from waste and dairy effluent disposal sites. The gas phase may include carbon dioxide, methane, nitrogen, ammonia and minor odoriferous gases, such as volatile fatty

Fig. 6 Gas (CO_2 and CH_4) forming a high bicarbonate froth from dairy whey disposal in a limestone aquifer at Timboon (Shugg 1998)

acids, mercaptans and esters. The fermentation of organic material in the waste plume represents a micro-milieu of the wider geological spectrum. An example of the dynamics of gas generation is depicted in a case in waste disposal into porous limestone near Timboon in Victoria as Fig. 6, the fermentation and souring of the effluent generates gas and acidic conditions, which are highly reactive with the limestone. Very high bicarbonate alkalinity is generated under these conditions.

6 Chemical Evolution

Evidence of a wide distribution of the high bicarbonate water facies can be based on the inference and by examining the distribution of extreme bicarbonate values in groundwater analyses.

6.1 Geochemical Evolution of High Bicarbonate Groundwater

Interest in high bicarbonate groundwater began in antiquity and later in the association with hardness and scale formation and the impact on industrial age machines. The recognition of the geochemical processes commenced with the evolution of chemical science from the days of Priestly (1733–1804) and the discovery of carbon dioxide by Black (1728–1799). Discrete processes such as diagenesis, petrogenesis, lithogenesis and coalification can contribute to the bicarbonate buildup in pore fluids. The literature abounds with descriptions of bicarbonate water formation (Foster 1950). Empirical observations in the Great Artesian Basin of Australia by Chebotarev (1955) identified trends in the evolution and metamorphism of natural waters in extensive groundwater flow systems in which the development of a bicarbonate facies could be recognised. In North America rock, water reactions that led to the composition of perennial spring waters of the Sierra were described by Garrels and McKenzie (Garrels and McKenzie 1967). In this description the recognition of the evolution bicarbonate facies was made. Mass balance computations involving the weathering of silicate minerals in a contiguous flow system with isothermal evaporation suggested that the terminus of the flow systems would be marked by the formation of sodium bicarbonate brines. Champ et al. (1978) described predominance of different ion suites within redox zones in aquifers and differentiation in the thermodynamics associated with closed and open flow systems that could lead to specific anion speciation. It is axiomatic that in many instances groundwater composition can be accounted for by examination of the rock mass geochemistry, the dynamics of the flow system and hypotheses relating to composition can be tested and modeled (Appelo 1994). Appelo (1994) identified that cation exchange and carbonate reactions in flushing aquifers can result in the formation of high bicarbonate waters.

In the Pocheon spa-land area of South Korea, Chae et al. (2006) illustrated that the formation of bicarbonate waters could be modeled using a mass balance and mixing approach. Chemical engineers have synthesized the formation of soda ash (sodium carbonate) and baking soda–sodium bicarbonate in the Solvay and Leblanc processes using salt and limestone.

6.2 Chemical Evolution and Development of High Bicarbonate Facies in Victoria

Locally in Victoria the development of a high bicarbonate groundwater facies was recognised as a phenomenon in many groundwater flow systems (Newbery 1867; Johns 1968, 1971; Blake 1980, 1989; Shugg and Knight 1994). Accession of the bicarbonate ion in groundwater can be explained by simple geochemical reactions. These may involve the hydrolysis of silicate minerals, carbonate solution—precipitation, oxidation of organic matter, sulphide oxidation and reduction, cation exchange and clay mineral reactions. Blake (1980) illustrated how the sodium bicarbonate water could be developed with carbonate solution driven by the hydrogen ion production from and clay mineral reactions involving cation exchange. In anthropogenic settings, high alkalinity—bicarbonate waters accompany the fermentation and biological mediated oxidation of organic matter and is often observed associated with waste disposal operations (Shugg 1998).

6.3 Rapidity of Development of the Sodium Bicarbonate Facies

Local examples of the rapidity of the development of the bicarbonate facies occur in the bedrock flow systems developed near the Dandenong Ranges (Shugg 1996) and in the deltaic sediments of the Snowy River at Orbost (Sinclair Knight Merz 2000). In each instance the high bicarbonate levels occurred within the first few kilometers of the commencement of the flow system. In the Namoi River valley of New South Wales, Merrick et al. (1986) identified the development of sodium bicarbonate waters in the extensive lithic and cobble gravels attributing the bicarbonate water formation to silicate hydrolysis, cation exchange and carbonate solution—precipitation reactions.

6.4 Carbonated Class of High Bicarbonate Groundwater in Victoria

In Central Victoria the high bicarbonate cold carbonated mineral occurs in distinct flow systems that are litho—structurally controlled. Detailed underground mapping of the gold fields (Whitelaw and Baragwanath 1923) provided evidence of the repetition, bifurcation, coalesce and linear persistence of many of the fissure networks. The fissure networks were mapped over many kilometers in the mine workings and on the surface by early geological mappers in the 1850–1860s. In Central Victoria the cold carbonated mineral waters have bicarbonate concentrations between 1,500 and 2,000 mg/L and a chloride concentration of 70–350 mg/L. Sodium is the dominant cation, although magnesium may be co-dominant. There are many other high bicarbonate water associations including non-gassy varieties and some with extremely high sulphate concentrations.

7 Conclusions

Examination of groundwater records has enabled the identification of the statistical abundance of the high bicarbonate groundwater in Victoria. The concentrations of bicarbonate form a statistical continuum. Bicarbonate is the second most abundant anion with 6 % of bores recording greater than 800 mg/L bicarbonate (as HCO_3^-). By comparison, chloride is the most abundant anion and occurs at levels of more than 3,500 mg/L for the same percentile.

Hydrogeologically, the most common location for high bicarbonate occurrence is identified as aquifer zones where mixing or flushing occurs, zones in the evolution of water types in deep flow systems, ascension zones and in chromatographically active redox dynamic plumes emanating from waste disposal operations.

Correlation with particular rock types suggests that lithic and labile sands and sandstones develop waters with the highest bicarbonate concentrations while coaly or ligneous sands may also possess very high concentrations. Changes in bicarbonate depth relationships of groundwater in Victoria partly reflect the sample group, and a clustering around the modal concentration of nearly 350 mg/L. Deep groundwater with extreme bicarbonate concentrations exists in regional groundwater flow systems in bedrock aquifers and in the deep sedimentary basins.

In the various hydrogeological settings, the key factors influencing the development of the high bicarbonate—sodium bicarbonate facies relate to the role of proton exchange (clay reactions), ion exchange, carbonate solution, sulphate reduction and bacterially mediated oxidization of organic matter. Local variations in water composition may relate to flow dynamics, pH control and redox. In sedimentary basins and alluvial aquifers a clear trend to the development of a sodium bicarbonate water exists in "dirty formations" which possess carbonaceous and carbonate rich beds in sandy sequences.

Bedrock systems exhibit the development of high bicarbonate waters where mixing or reflux is a process. The role of geological process has been identified in the recovery of the water composition after the dewatered aquifers of the Central Victorian gold fields. Mining induced recharge by fresh water and resulted in the loss of the bicarbonate levels. After mining ceased and many decades later the groundwater systems of the gold fields recovered and re-established the high bicarbonate facies of the deep groundwater.

Anthropogenic sources associated with waste disposal demonstrate the role of fermentation and reduction of organic compounds in developing high bicarbonate groundwater. The plumes also illustrate variation in processes with changes in composition associated with redox and chromatographic effects.

References

Appelo CAJ (1994) Cation and proton exchange, pH variations and carbonate reactions in a freshening aquifer. Water Resour Res 30(10):2793–2805

Blake R (1980) Geology and hydrogeology of the early Tertiary sediments of the Otway Basin. M.Sc. thesis, Latrobe University

Blake R (1989) The origin of the high sodium bicarbonate waters in the Otway Basin, Victoria, Australia. In: Miles DL (ed) Water-rock interaction. Balkema, Rotterdam, pp 83–85

Chae G-T, Yun S-T, Kim K, Mayer B (2006) Hydrogeology of sodium—bicarbonate type bedrock groundwater in the Pocheon spa area, South Korea: water rock interaction and hydrologic mixing. J Hydrol 321:326–343

Champ DR, Gulens J, Jackson ER (1978) Oxidation reduction sequences in groundwater flow systems. Can J Earth Sci 16:12–23

Chebotarev II (1955) Metamorphism of natural waters in the crust of weathering. Geochemica et Cosmochimica Acta 8:22–48, 137–170, 198–211

Forbes CF, Showers JAF (1988) Mine dewatering and water disposal Bendigo goldfield. In: The third international mine water conference, Melbourne, Australia, Oct 1988, pp 415–424

Foster MD (1950) The origin of high sodium bicarbonate in the Atlantic Gulf Coastal Plain. Geochim Cosmochim Acta 27:205–214

Garrels RM, McKenzie FT (1967) Origin of the chemical composition of some springs and lakes. Am Chem Soc Adv Chem Ser 67:222–242

Johns MW (1968) Geochemistry of groundwater from upper Cretaceous—lower Tertiary sand aquifers in South Western Victoria, Australia. J Hydrol 6(4):337–531

Johns MW (1971) Geochemistry of groundwater in the Otway Basin of South Western in Victoria. In: Wopfner H, Douglas JG (eds) The Otway Basin of South Eastern Australia (Spec. Bull., geological surveys of South Australia and Victoria), pp 363–375

Maund J (1855) On the mineral waters of Victoria. Trans Proc Victorian Inst (Philosophical) 70–74 (referred to by Taylor 1894, Geol Surv Prog Rep 8:19)

Merrick NP, Ross JB, Williams RM (1986) Groundwater in the lower Namoi Valley, New South Wales. In: Proceedings of the international conference on groundwater systems under stress. Australian Water Resources Council Conference Series, vol 13, pp 461–471

Merz SK (2000) PAV determination for the Curlip Gravel and Jarrahmond formation aquifers. Report to the Department of Natural Resources and Environment

Newbery JC (1867) Mineral waters of Victoria. Trans R Soc Vict 8:278–283

Shugg A (1996) Hydrogeology of the Dandenong Ranges fractured rock aquifers and the comparison with similar aquifers in Victoria. National Centre for Groundwater Management, University of Technology Sydney, 850 p (M.Sc. thesis)

Shugg A (1998) Whey disposal Allansford Western Victoria. In: Weaver TR, Lawrence CR (eds) Proceedings of the IAH international groundwater conference. Groundwater: sustainable solutions, Melbourne, Australia, 8–13 Feb 1998, pp 563–568

Shugg A, Knight M (1994) Hydrogeology of the mineral springs of Central Victoria. In: Proceedings of the water down under conference, 21–25 Nov 1994. International Association of Hydrogeologists and the Institute of Engineers, Adelaide, Australia

Whitelaw HS, Baragwanath W (1923) The Daylesford Goldfield. Geological survey of Victoria. Bulletin 42

Industrially Valuable Components in Oilfield Waters of Russia

Arkady Voronov and Natalia Vinograd

Abstract A large volume of groundwater is extracted from oilfields, along with oil. Therefore, the processing of oilfield water can be commercially useful. Unfortunately, in most cases oilfield water is not in use; a great amount of it disposes onto land surface that causes major environmental problems. Bromine and iodine are the most prospective for extraction from oilfield waters. Bromine waters widely occur within oilfields of East-European and Siberian platforms. Oilfield waters often contain much iodine. There are some regularities of iodine accumulation in groundwater. Boron, magnesium, lithium and strontium can also be found in oilfield water in commercial concentrations. In Russia, the problem of extraction of industrially valuable components from groundwater has been investigated for a long time. Most attention was paid to bromine and iodine extraction. In Russia, various hydrogeological structures with common regularities of industrially valuable groundwater prevalence can be combined into six industrially valuable water provinces according to geological and hydrogeological criteria. Great reserves and a variety of industrially valuable waters in Russia allow extraction of a great number of useful components. It is necessary to develop technologies of the extraction of lithium, strontium, copper, zinc, rubidium, cesium, potassium, magnesium and other components from oilfield water.

Keywords Oilfield water · Deep groundwater · Industrially valuable components · Provinces of industrially valuable waters

---

Arkady Voronov: deceased.

A. Voronov · N. Vinograd (✉)
Department of Hydrogeology, Intstitute of Earth Science, University of St. Petersburg, Universitetskaya nab. 7/9, St. Petersburg 199034, Russia
e-mail: nv.70@hotmail.com; n.vinograd@spbu.ru

W. Balderer et al. (eds.), *Thermal and Mineral Waters*, Environmental Earth Sciences, DOI: 10.1007/978-3-642-28824-1_10, © Springer-Verlag Berlin Heidelberg 2014

1 Introduction

A large volume of groundwater is extracted from oilfields, along with oil. The longer the oilfield exploitation, the greater the share of water in the extracted fluid (in some cases, it is more than 50 %). The greatest share of water is observed when the full watering of the well occurs.

As a rule, oilfield water comes from great depths; therefore, it is highly mineralized and has high concentration of many useful components. Therefore, the processing of oilfield water can be commercially useful. In this case the interests of oil producers and hydrogeologists are diverse. However, processing of oilfield water can be profitable for both parties, as the value of components extracted from oilfield waters can even exceed the oil value. Unfortunately, in most cases, oilfield water is not in use; a great amount of it disposes onto the land surface which causes major environmental problems.

2 Discussion

2.1 Bromine and Iodine in Oilfield Water

Bromine has the highest concentrations in water of all microcomponents. Its content in brines can be up to 6–7 g/l. When bromine concentrations exceed 250 mg/l, its extraction from oilfield water is considered profitable (Bondarenko and Kulikov 1984). As a rule, bromine contents rise with mineralization and metamorphization values (Fig. 1).

Bromine is a highly valuable component. Bromides are 95 % soluble in water. The greatest amount of bromides is accumulated in seawater, and the main sources of bromine in groundwater are buried seawater and dissolution of halogen rocks.

Bromine waters widely occur within oilfields of East-European and Siberian platforms (Yazvin 1977). Bromide-rich brines, in the north of the East-European platform, in Paleozoic rocks, contain 375–900 mg/l of bromine; and their total dissolved solids (TDS) are up to 190 g/l.

In Southern Timan (Pechora artesian basin), in Cambrian-Paleocene rocks, brines with TDS of 50–235 g/l and bromine content of up to 800 mg/l, are discovered.

In Volga region, bromine-rich water is widespread. In the area of Perm region up to the eastern border of Pre-Ural depression, highly saturated bromine-rich brines occur under the anhydrite Permian sediments. They contain up to 1.8 g/l of bromine (Krasnokamsk).

Within the Siberian Platform, in deep aquifers of Angaro-Lensky and Yakutsky artesian basins, there are highly saturated bromine-rich brines, which contain up to 7 g/l of bromine. It is very probable that the waters of new oilfields found in aquatic parts of ancient platforms contain much bromine.

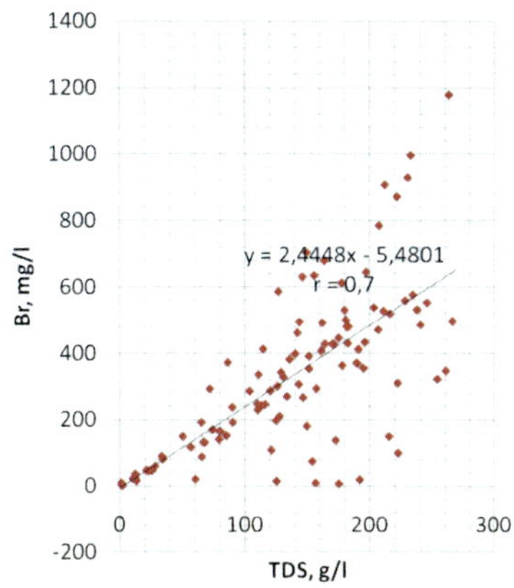

Fig. 1 Relationship between bromine contents and TDS in groundwater of the Northern part of East-European Platform, (216 samples, all aquifers)

In Russia, more than 70 % of bromine is extracted from groundwater. The other 30 % is extracted from surface brines of lakes, sea gulfs and disposals of potassium manufactures. The main source of bromine is brines of Krasnokamsk occurrence in Perm region. It is not related to oilfields; the wells were purposely drilled for a bromine plant. Groundwater of chloride-calcium-sodium composition is used for bromine production.

Iodine is another element extracted from groundwater. Only water and aquatic plants are sources of iodine, as it is not accumulated in solid rocks. As a rule, waters of medium and low mineralization contain iodine. Accumulation of iodine occurs in water with higher content of organic matter. The main iodine concentrators are marine plants and organisms. Iodine in iodide form predominates in the plants. As the algae sedimentation occurs mostly in freshened sea water, high iodide contents are observed in the sedimentation waters with lower mineralization. Oilfield waters often contain much iodine. The interaction between rocks and water occurs with the presence of organic matter, which defines concentrations and migration forms of iodine in groundwater. The main factors that cause the transport of iodine to groundwater are reduction conditions, alkaline media and temperature.

There are some regularities of iodine accumulation in groundwater (Kudelsky 1978): iodine accumulates in the thick (1,000–1,500 m) layers of buried marine sediments enriched in organic matter, the total thickness of sedimentary cover should be about 3,000–5,000 m and more. Oil and gas basins occurring in the deep geological structures often contain iodine-rich waters. Iodine contents in groundwater also depend on the age of water-bearing rocks. For example, in Mezo-Cenozoic basins, the average iodine content in groundwater is 36.3 mg/l; however, in Paleozoic basins it is only 12.5 mg/l.

The composition of iodine-rich waters is mostly chloride-bicarbonate or bicarbonate-chloride sodium.

In Russia, iodine is manufactured in Slavyano-Troitsky plant in Krasnodar Region. In Western Siberia, the new Tumen' occurrence of iodine-rich water is prospected. The reserves of this occurrence are great. A plant constructed here could produce three times more iodine than is produced in Russia now. The water with TDS of about 70 g/l contains 25–45 mg/l of iodine. The productive Neocom aquifer (K_1) occupies the northern part of Tobol artesian basin.

2.2 Boron

Boron in groundwater has a different origin. Oil- and gas-fields boron occurs in groundwater due to the decay of organic matter. Boron contents in groundwater can reach n·10 – n·100 mg/l (Krainov et al. 2004).

Boron-rich waters can be of different chemical composition. In Russia, boron content of more than 200 mg/l is considered industrially valuable; 24.4 % of world boron resources are situated in Russia (Petrov et al. 2011). About 50 % of world boron reserves occur in groundwater, and 20–30 % of world boron production is extracted from groundwater (USA, Italy, etc.)

2.3 Magnesium

Magnesium occurs in groundwater in significant concentration, as it is macro-component. About 25 % of this element is extracted from water. It is commercially effective to produce magnesium from seawater. USA is a world leader in magnesium production. In this country, there is only one plant, where magnesium is produced from solid rocks. Other plants use seawater or groundwater. In Israel, magnesium is produced from Dead Sea brines.

2.4 Lithium

About 55–63 % of world lithium reserves occur in mineralized waters; 30 % of the world lithium production is extracted from liquid raw material. The most advanced technologies on lithium extraction are developed in the USA. In Michigan and Oklahoma, there are oilfield waters that contain up to 3 g/l of lithium. In Russia, oilfield waters have much poorer lithium contents than in the USA; however, it can be commercially effective to extract lithium from waters, where its content is more than 10 mg/l. In South America, lithium is extracted from surface and subsurface brines of salars (salt lakes). The largest lithium occurrence in the world is Salar de Uyuni (Bolivia), where about 57 % of world reserves occur. Lithium occurs also in salars of Chili and Argentina.

2.5 Strontium

Strontium is traditionally extracted from strontium-rich minerals; however, at present there are technologies of its extraction from groundwater, as about 24 % of world strontium reserves are in groundwater.

3 Provinces of Commercially Valuable Waters

Within the area of the former Soviet Union, the presence of industrially valuable water and prospects of its possible utilization were assessed (Fig. 2; Bondarenko et al. 1986). Various hydrogeological structures with common regularities of industrially valuable groundwater prevalence were systemized into industrially valuable water provinces. There are main criteria in distinguishing them: (1) geological and tectonic setting of the region, (2) age and lithological and facial composition of host rocks, (3) conditions of industrially valuable water occurrence, and (4) mineralization, chemical composition and concentration of micro-components in water. Considering these factors, the next provinces of industrially valuable waters are distinguished in Russia (Bondarenko and Kulikov 1984):

The Province of East-European platform has a thick zone of industrially valuable waters (from 1 up to 5 km), occurring in depressions of Precambrian crystalline basement, which is characterized by stagnant hydrogeological regime. Main reserves of oilfield industrially valuable waters are associated with terrestrial and carbonate rocks of Devonian, Carboniferous and Permian age. In Lower Permian and Devonian rocks there are salt layers.

Mineralization of brines rises with depth from 36 to 430 g/l; the chemical composition changes from chloride sodium to chloride sodium-calcium. Bromine concentration reaches 3.5–6.3 g/l; potassium—13–15 g/l; iodine—50–90 mg/l; boron—0.5–1 g/l; strontium—0.9–5.0 g/l; rubidium—2–3 mg/l; cesium—1–2 mg/l.

In *Pre-Caspian province* the crystalline basement occurs on the depth of 15–20 km. The sedimentary cover is of Lower-Proterozoic, Paleozoic, Mesozoic and Cenozoic age. Salt layers of Permian Kungur stage widely occur. In overlaying rocks, chloride-sodium brines with low concentration of microcomponents prevail. Within salt layers, interlaying dolomites and anhydrites contain chloride-sodium, chloride sodium-magnesium and chloride sodium-magnesium-potassium concentrated brines (TDS 320–520 g/l) with high contents of potassium (up to 60 g/l), bromine (up to 9.3 g/l), strontium (up to 3.5 g/l) and boron (up to 1 g/l). However, reserves of these intersalty waters are relatively small. The mentioned elements in commercially valuable concentrations are present also in underlying Paleozoic aquifers.

The Province of Siberian Platform is a large tectonic depression with a depth of up to 5 km and more. The sedimentary cover consists of terrestrial, carbonate,

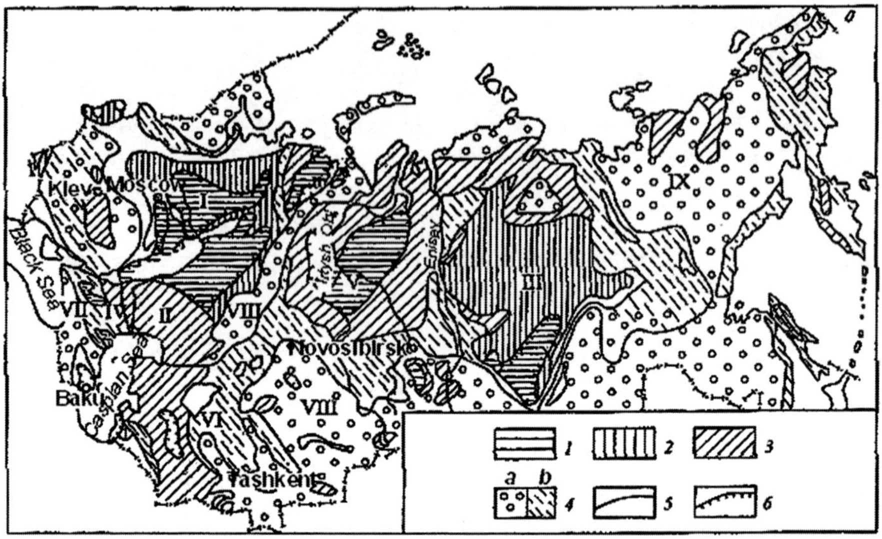

Fig. 2 Scheme of the industrially valuable water provinces of the former Soviet Union (Bondarenko et al. 1986). *Legend* Provinces of platforms: *I*—East-European (Russian); *II*—Pre-Caspian; *III*—Siberian; *IV*—Scythian; *V*—West-Siberian; *VI*—Turanian. Provinces of folded areas: *VII*—Alpine; *VIII*—Hercynian; *IX*—Mesozoic; *X*—Cenozoic. Areas of industrially valuable waters: *1*—very prospective; *2*—prospective; *3*—less prospective: *4*—unprospective (*a*—folded areas and crystalline shields; *b*—platforms). Borders: *5*—of provinces; *6*—of groundwater occurrences

halogenic and volcanic rocks of various age (from Late Proterozoic to Cenozoic). The platform has some peculiarities: (1) traprock magmatism in Late Proterozoic, Devonian, Early Permian and Triassic periods; (2) wide occurrence of permafrost rocks up to 1 km thick (Vilyui Sincline); (3) intensive disjunctive tectonics having caused the discharge of brines to the land surface; and (4) presence of Lower-Cambrian and Devonian salt layers.

Low-Cambrian halogen sulfate-carbonate rocks have a thickness of up to 3 km. They occur in the southern and southwestern part of the region. Devonian salts and gypsums (0.3–0.7 km thick) occur in the north and east.

The thickness of the zone of brines exceeds 5 km. Their total geological reserves are estimated as $1.5 \cdot 10^{14}$ tons. Chloride sodium-calcium brines occurring under the salt layers have the highest TDS (400–640 g/l); TDS of chloride sodium-calcium brines in underlying aquifers is 160–400 g/l. The specific feature of the brines of Siberian Platform is unique high concentration of industrially valuable components—up to 8.8 g/l of bromine, up to 7 g/l of iodine, up to 35 g/l of potassium, up to 0.06 g/l of rubidium.

The Province of Scythian plate embraces the area of Tersko-Kumsky and Azovo-Kubansky artesian basins. The Paleozoic crystalline basement occurs on the depth of more than 3–5 km. The basement rocks are intensively dislocated and metamorphized. Mesozoic-Cenozoic sedimentary cover consists of terrestrial and terrestrial-carbonate rocks.

The groundwater temperature on the depth of 1 km is 60–80 °C, on the depth of 3 km—140–170 °C. Lithogenous hydrogeodynamic regime causes high hydraulic head of groundwater in Paleogene, Cretaceous and Jurassic aquifers. Thermal compressional regime is supposed to exist on the depths of 5–6 km.

Brines predominate in hydrogeological section of the whole area. They contain up to 920 mg/l of bromine, 5–50 mg/l of iodine, 10–90 mg/l of boron, up to 200 mg/l of potassium, up to 300 mg/l of strontium and up to 2 mg/l of rubidium. In Jurassic aquifer of East-Kuban depression there are chloride sodium and sodium-calcium brines of very high TDS (200–435 g/l) with 500–600 mg/l of bromine, 50 mg/l of iodine, 2.5 g/l of potassium, 1.6 g/l of strontium.

The Province of West Siberian plate occupies the area of West Siberian artesian basin, which is subdivided into smaller basins. Meso-Cenozoic polyfacial sandy clays compose sedimentary cover. Carbonate and halogen rocks are absent. Folded Paleozoic basement is broken by deep faults, where carbon dioxide fluids penetrate to sedimentary cover. Cretaceous and Jurassic aquifers are of commercial interest. Methane chloride sodium brines and bicarbonate-sodium and chloride calcium brackish water (TDS up to 36 g/l) contain 1.55 mg/l of iodine, 25–160 mg/l of bromine, up to 200 mg/l of potassium, 300 mg/l of strontium, 2 mg/l of rubidium.

Jurassic aquifer contains waters with TDS of 65 g/l. Their specific feature is higher potassium (up to 1 g/l), strontium (up to 0.8 g/l) and bromine (about 200 mg/l) contents; meanwhile, iodine contents are rather low (11 mg/l and less).

In *Sakhalin Province*, there are several small artesian basins containing iodine- and bromine-rich waters. The basins are formed by terrestrial Neogene rocks up to 8 km thick. The investigated depth is 2.5–3.2 km. In this part of the section, there are fresh bicarbonate, salty bicarbonate-chloride and salty chloride sodium waters with TDS of up to 36 g/l. The iodine content is about 10 mg/l, bromine content is 90 mg/l.

4 Conclusion

In Russia, the problem of extraction of industrially valuable components from groundwater is investigated for a long time. Particularly, problems of extraction technologies, perspective geological structures for the search of such waters and commercial effectiveness were analyzed. Most attention was paid to bromine and iodine extraction; however, geological and hydrogeological criteria of the presence of other industrially valuable components and possible technologies of their extraction were analyzed.

Technologies of extraction of lithium, strontium, copper, zinc, rubidium, cesium, potassium and magnesium and other components from waters of different chemical composition are developed.

References

Bondarenko SS, Kulikov GV (1984) Underground industrially valuable waters. Nedra, Moscow, p 358 (in Russian)

Bondarenko SS, Vartanyan GS et al (1986) Methods of research and assessment of the resources of deep groundwater. Nedra, Moscow, p 479 (in Russian)

Krainov SR, Ryzhenko BN, Shvets VM (2004) Geochemistry of groundwater. Nauka, Moscow, p 677 (in Russian)

Kudelsky AV (1978) Forecasting of the regional iodine contents in groundwater. Nauka i tekhnika, Minsk (in Russian)

Petrov SV, Polekhovsky YuS, Borozdin AP et al (2011) Resources of mineral raw materials of the world. St. Petersburg University, St. Petersburg (In Russian)

Yazvin LS (ed) (1977) Hydrogeology of the USSR. Nedra, Moscow (in Russian)

Appendix A
Geographic and Geologic Terms Connected With Egypt

Egypt Official name is the Arab Republic of Egypt.

Total area is 1,002,450 km^2, which makes Egypt the 29th largest country in the world. Only 0.632 % of the area is made up of water bodies.

It is located in the northeast corner of the African continent, on the Mediterranean Sea, at a crossroad between Africa, Asia and Europe. It is bordered to the east by the Red Sea, Palestine and Israel to the northeast, Libya to the west, and Sudan to the south.

Population is about 84.9 million, which makes Egypt the most populated country in the Middle East, third in the African continent, and fifteenth in the world (Central Agency for Public Mobilization and Statistics of the Arab Republic of Egypt 2013). The majority of the people live along the banks of the Nile River and its deltas, where the only arable land is found. It is estimated that about 99 % of the population uses only about 5.5 % of the land (Hamza 2011).

Administration: Egypt is divided into 28 independent administrative units, the so-called governorates.

Climate is hot and desert, with dry summers and moderate winters. Two seasons can be distinguished: a mild winter from November to April, and a hot summer from May to October. Temperatures vary widely in the inland desert areas: in summer they may range from 7 °C at night to 43 °C during the day; in winter they can be as low as 0 °C at night and as high as 18 °C during the day.

Geology: Four main geological areas can be distinguished in Egypt, namely: (1) Nile River valley and delta, (2) Western Desert, (3) Eastern Desert, (4) Sinai Peninsula (Biswas et al. 2008). The lowest point is Qattara Depression: −133 m b.s.l.; the highest point is Mount Catherine: 2.629 m a.s.l.

Hydrogeology: The main source of fresh water is the Nile River. The river supplies 56.8 billion m^3 of freshwater every year, which represents 97 % of all renewable water resources in Egypt. Groundwater for drinking and irrigation purposes is extracted also from the Nubian Sandstone aquifer, the Nile aquifer, the Moghra aquifer and the Coastal aquifer (Biswas et al. 2008).

W. Balderer et al. (eds.), *Thermal and Mineral Waters*, Environmental Earth Sciences,
DOI: 10.1007/978-3-642-28824-1,

Sahara Is the largest desert zone on the Earth, located in North Africa, with a total area of about 9,064,300 km^2. It stretches about 5,700 km from the Red Sea on the East to the Atlantic Ocean on the West, and about 1,800 km from the Mediterranean coasts on the North to the Sahel region (i.e., belt of the semi-arid tropical savanna) and valley of the Niger River on the South. The Sahara covers large parts of 11 countries: Egypt, Libya, Tunisia, Algeria, Morocco, Sudan, Chad, Niger, Mali, Mauretania, Western Sahara. The surface of the desert consists of large areas of sand dunes (erg), stone plateaus (hamadas), gravel plains (reg), dry valleys (wadis), and salt flats (Cloudsley-Thompson 1984).

The mean annual temperatures in the Sahara exceed 30 °C. In the hottest months, temperatures can rise over 50 °C, and can fall below freezing in the winter. Usually the annual rainfall is less than 25 mm.

Most of the rivers and streams are seasonal or intermittent, except the Nile River. Underground aquifers, which sometimes reach the surface, form oases, e.g. the Bahariya, Siwa, Farafra.

The highest point is Emi Koussi Mountain in the Tibesti Mountains, central Sahara. The mountain reaches 3415 m a.s.l, and geologically is considered as an example of a pyroclastic shield volcano.

The lowest point is Qattara Depression, −133 m b.s.l., which is located in the northwest part of Egypt, in the Matruh Governorate, and covers about 19,605 km^3. The surface of the depression is covered by salt pans, sand dunes and salt marshes.

Western Desert This geological region of Egypt covers about two-thirds of the country's territory. It stretches from the Nile Valley in the east to the Egypt-Libyan border in the west, and from the Mediterranean coast in the north to the Egypt-Sudan border in the south. It is one of the most arid regions in Egypt. However, the most well-known Egyptian oases like Siwa, Bahariya, Farafra, Kharga and Dakhla occur in this region.

The Egyptian Western Desert belongs to the Lybian Desert, which is one of the 14th major regions of the Sahara.

Eastern Desert Geological region of Egypt is one of the regions of the Sahara desert. It is located east of the Nile River, between the Nile Valley and the Red Sea and Gulf of Suez. It extends from Lake Manzala on the north to Sudanese border on the south, and further to Eritrea. The main geographical feature of this region is the Eastern Desert mountain range that runs along the Red Sea coast, with the highest peak, Shaiyb al-Banat (2187 m a.s.l). This region is one of the richest in natural resources like oil and coal.

White Desert Is part of the Egyptian Western Desert and is located about 45 km northeast of Qasr el-Farafra. The area is renowned for its spectacular scenery. The desert is composed of massive chalk rock formations and has a white, chalk-white and cream color. The landscape constitutes strewn massive chalk rocks and boulders of different, alien shapes. Such a landscape was formed by wind and sand erosion over the broken-down chalk plateau, leaving the harder rock shapes standing while the softer parts eroded away. The White Desert is considered as the natural wonder of Egypt, and is now a protectorate, known as the White Desert Park.

Black Desert Is part of the Egyptian Western Desert, located approximately 50 km to the south of Bawiti, a little to the north and north-east of the White Desert. The surface of this part of the desert is covered by eroded volcanic material in the form of rocks and gravels of brown-orange and black color. The majority of these dark rocks are composed of diabase. The characteristic feature of the landscape here are hills and single volcano-shaped mountains with a huge amount of black and brawn-colored stones in the upper parts and on the slopes between beige sands. The highest peak is the Gebel el-Ingiz.

Sinai Is a triangular-shaped peninsula, the only part of Egyptian territory located in Asia. It is linked by its northwestern corner to Egypt's mainland, at the Gulf of Suez. The peninsula is located between the Mediterranean Sea on the north and the Red Sea on the south, the Gulf of Suez on the west, and the Gulf of Aqaba on the east. The total area is about 61,200 km^2 and the population is approximately 0.5 million people. The majority of the Sinai Peninsula is a desert area. Geographically the northern part forms highlands and the southern parts are mountains, with the highest peak the Mount of St. Catherine, 2,640 m a.s.l.

This region is considered to be an active seismic area due to the presence of the triple junction of the Gulf of Suez and Aqaba and the Red Sea.

The geology of the Sinai Peninsula ranges from the Precambrian basement rocks to Quaternary deposits. The Quaternary deposits cover the northern part and along the Gulf of Sinai and the Mediterranean Sea coast. The Mesozoic limestone covers the wide area from the central part of Sinai Peninsula, while the Precambrian rocks outcrop and cover a wide area in the southern part (Rabel 2011).

Appendix B
Photo Captions

Photos 1, 2, 3, 4, 5, 6, 7, 8, 9, 10, 11, 12, 13

Photo 1 River Nile and centre of Cairo in evening (photo by A. Porowski)

W. Balderer et al. (eds.), *Thermal and Mineral Waters*, Environmental Earth Sciences,
DOI: 10.1007/978-3-642-28824-1, © Springer-Verlag Berlin Heidelberg 2014

Photo 2 **a**, **b** Scientific session and the meeting of the IAH Commission on Mineral and Thermal Water (CMTW) in Cairo. The meeting was organized together with the Egyptian Society of Engineers. Local media was present and the information about the meeting as well as a few interviews were given by the members who appeared on Egyptian television. **b** Eng. Hussein Idris being interviewed

Photo 3 **a**, **b** Qarun Petroleum Company, Qarun oil field, Western Desert, Egypt (photo by A. Porowski)

Photo 4 Landscape of the so-called Black Desert, part of the Western Desert, Egyptian Sahara (photo by A. Porowski)

Photo 5 Landscape and erosion remnants characteristic for the so-called White Desert, part of the Western Desert, Egyptian Sahara (photo by A. Porowski and E. Kayukova)

Photo 6 **a**, **b** Bahariya Oasis, Western Desert, Egyptian Sahara (photo by A. Porowski)

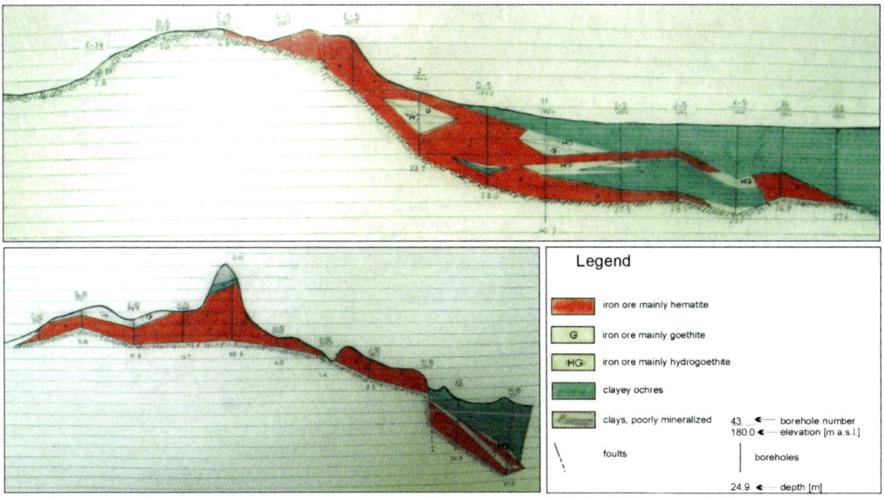

Photo 7 Exemplary sketch of the geological settings in the area of iron ore extraction. The iron ore mine in Bahariya Oasis is the largest in Egypt (compiled by A. Porowski based on the data from iron ore mine)

Photo 8 Groundwater intake for irrigation purposes in Bahariya Oasis

Photo 9 General view on the area of Oyun Musa (Moses Springs), Sinai Peninsula. The depth to the aquifer in Nubian formation here is about 30 m, which were allowed to be dug to reach drinking water (photo by A. Porowski)

(a)
13/10/09 14:20
(b)

Photo 10 **a**, **b**, **c** Ras Badrun oil field, Sinai Peninsula: old oil-well with artesian flow of thermal water. Outflow temperature measured by Dr. Balderer was 64.9 °C (photo 10b)

(a)
(b)

Photo 11 **a**, **b**, **c** Walking trip to Mount Sinai (photo by A. Porowski and E. Kayukova)

Photo 12 View on the oil platform during sunset over Suez Canal, Sinai Peninsula, Egypt (photo by A. Porowski)

Photo 13 Eng. Hussein Idris, main organizer of the 38th CMTW IAH meeting in Cairo, Egypt, 2009. He passed away in January 2013. This outstanding engineer and hydrogeologist was a member of the CMTW since its establishment. He served for many years as vice chairman of the Commission, a great colleague with unforgettable personality; a teacher for many of us, who will stay in our minds forever (photo by A. Porowski)

References

Biswas AK, Rached E, Tortajada C (eds) (2008) Water as a human right for the Middle East and Africa. Routledge/IDRC, p 192

Central Agency for Public Mobilization and Statistics of the Arab Republic of Egypt (Aug 2013)

Cloudsley-Thompson JL (1984) Sahara desert. Pergamon Press, Oxford

Hamza W (2011) Land use and coastal management in the third countries: Egypt as a case (unpublished; www.oceandocs.org)

Rabel T (2011) Tectonic model of the Sinai Peninsula based on the geophysical investigations. In: Closson D (ed) Tectonics. (intechopen.com)

Printed by Publishers' Graphics LLC
MLSI140321.15.19.24